The third edition of *Experiments in Plant Tissue Culture* makes available new and updated information that has resulted from recent advances in the applications of plant tissue culture techniques to agriculture and industry. This laboratory text features an enlarged section on terminology and definitions as well as additional information on equipment and facilities.

This comprehensive text takes the reader through a graded series of experimental protocols and also provides a preliminary review of each topic. After an introductory historical background, there are discussions of a plant tissue culture laboratory, aseptic techniques, and nutritional components of media. Subsequent chapters are devoted to callus induction, organ formation, cell suspensions, somatic embryogenesis, root cultures, micropropagation, anther and pollen cultures, xylem cell differentiation, isolation and fusion of protoplasts, storage of genetic resources, secondary metabolite production, and quantitation of procedures. A glossary of terms, a list of commercial sources of supplies, and a table of media formulations are included.

This book offers all the basic experimental methods for the major research areas of plant tissue culture, and will be invaluable to undergraduates and research investigators in the plant sciences. It will also be useful to researchers in related areas such as forestry, agronomy, and horticulture, and to those in commercial houses and pharmaceutical companies interested in acquiring the laboratory techniques.

*Experiments in Plant Tissue Culture*
Third Edition

This volume is dedicated to Professor Gautheret for his outstanding contributions to research and teaching of plant cell and tissue culture.

# EXPERIMENTS IN
# *Plant Tissue Culture*

Third Edition

**JOHN H. DODDS**
Michigan State University

AND

**LORIN W. ROBERTS**
University of Idaho

Published by the Press Syndicate of the University of Cambridge
The Pitt Building, Trumpington Street, Cambridge CB2 1RP
40 West 20th Street, New York, NY 10011-4211, USA
10 Stamford Road, Oakleigh, Melbourne 3166, Australia

First edition © Cambridge University Press 1982
Second edition © Cambridge University Press 1985
Third edition © Cambridge University Press 1995

First published 1995

*Library of Congress Cataloging-in-Publication Data*
Dodds, John H.
Experiments in plant tissue culture / John H. Dodds and Lorin W. Roberts – 3rd ed.
p. cm.
Includes indexes.
ISBN 0-521-47313-6 (hardback). – ISBN 0-521-47892-8 (paperback)
1. Plant tissue culture – Experiments. I. Roberts, Lorin W. II. Title
QK725.D63 1995
581′.0724 – dc20         94–22943
CIP

A catalog record for this book is available from the British Library

ISBN 0-521-47313-6 hardback
ISBN 0-521-47892-8 paperback

Transferred to digital printing 2004

# Contents

| | |
|---|---|
| *List of tables* | *page* vii |
| *Foreword, by J. Heslop-Harrison* | ix |
| *Preface to the third edition* | xi |
| *Preface to the second edition* | xii |
| *Preface to the first edition* | xiii |

## I Background information

| | | |
|---|---|---|
| 1 | Culture of plant cells, tissues, and organs | 3 |
| 2 | Laboratory facilities | 19 |
| 3 | Aseptic techniques | 25 |
| 4 | Media composition and preparation | 42 |

## II Experimental: Callus and callus-derived systems

| | | |
|---|---|---|
| 5 | Initiation and maintenance of callus | 69 |
| 6 | Organogenesis | 82 |
| 7 | Cell suspensions | 92 |
| 8 | Somatic embryogenesis | 101 |

## III Experimental: Culture of organs and organized systems

| | | |
|---|---|---|
| 9 | Isolated roots | 117 |
| 10 | Micropropagation by bud proliferation | 126 |
| 11 | Anther and pollen cultures | 136 |

## IV Experimental: Isolated cells

| | | |
|---|---|---|
| 12 | Transdifferentiation of parenchyma cells to tracheary elements | 155 |

| | | |
|---|---|---|
| 13 | Isolation and culture of protoplasts | 167 |
| 14 | Protoplast fusion and somatic hybridization | 183 |

## V  Supplementary topics

| | | |
|---|---|---|
| 15 | Cryopreservation of germplasm | 195 |
| 16 | Production of secondary metabolites | 204 |
| 17 | Quantitation of procedures | 214 |

| | |
|---|---|
| *Abbreviations* | 229 |
| *Glossary* | 231 |
| *Commercial sources of supplies* | 235 |
| *Formulations of tissue culture media* | 240 |
| *Author index* | 243 |
| *Subject index* | 248 |

# Tables

| | | |
|---|---|---|
| 4.1 | Medium for *Nicotiana tabacum* stem callus | page 57 |
| 8.1 | Examples of plants in which somatic embryogenesis has been induced under in vitro conditions | 102 |
| 11.1 | A basal medium for the liquid culture of pollen | 145 |
| 13.1 | Some enzyme preparations exhibiting wall-degrading activity classified according to major function | 178 |
| 13.2 | Examples of combinations of enzyme mixtures used successfully for the preparation of protoplasts | 178 |
| 16.1 | Secondary metabolites produced in cell cultures at levels equal to or higher than those found in the intact plant | 206 |
| 17.1 | Various techniques for the measurement of growth and differentiation in plant tissue cultures | 215 |

# Foreword

The idea of experimenting with the tissues and organs of plants in isolation under controlled laboratory conditions arose during the latter part of the nineteenth century, finding its focus in the work of the great German plant physiologist Haberlandt some 80 years ago. Haberlandt's vision was of achieving continued cell division in explanted tissues on nutrient media – that is, of establishing true, potentially perpetual, tissue cultures. In this, he was himself unsuccessful, and some 35 years were to elapse before the goal was attained – as it could be only after the discovery of the auxins. Gautheret, Nobécourt, and White were the pioneers in this second phase. The research they set in train was at first mainly concerned with establishing the conditions in which cell division and growth would take place in explants, and in exploring the nutritional and hormonal requirements of the tissues; but this quickly gave place to a period during which cultured tissues were used as a research tool, in studying more general problems of plant cell physiology and biochemistry and the complex processes of differentiation and organogenesis. The achievements were considerable; but above all, the finding that whole plants could be regenerated from undifferentiated tissues – even single cells – in culture gave the method enormous power. In an extraordinary way this has meant that at one point in time the entity – a plant – can be handled like a microorganism and subjected to the rigorous procedures of molecular biology, and at another time be called almost magically back into existence as a free-living, macroscopic organism. The implications and applications of this finding are currently being explored in many contexts, not least in the field of practical application. If genetic engineering, involving the direct manipulation of the stuff of heredity, is ever to contribute to that part of human welfare that depends on the exploitation of plants, the procedures adopted will inevitably depend ultimately upon the recovery of "real" plants from cultured components. No wonder, then, that the technology has escaped from the confines of the university laboratory to become part of the armory of industry and agriculture!

Yet, notwithstanding the wide interest in the methods of plant tissue culture, the range of modern techniques has never hitherto been treated comprehensively in one text. Both novitiate and initiate have had to explore the large and scattered literature to unearth procedures appropriate to their interests. This volume makes good the deficiency, for the experiments described cover almost every aspect of the tissue culture art. However, this is far more than any cookery book. The methods of achieving growth, cell division, and morphogenesis in vitro are set in their appropriate contexts. The chapters not only describe how to carry out procedures, but offer lucid accounts of the historical background and interpretations of the results likely to be obtained, backed up by extensive bibliographies. The authors are peculiarly well fitted to have written such a text, with their extensive experience of the application, development, and teaching of tissue culture methods. Directed in the first instance toward students, their treatment of the topics will prove of immense value to a much wider range of readers, whatever their previous knowledge or field of potential application.

*Welsh Plant Breeding Station,*                                           J. Heslop-Harrison
*Plas Gogerddan, Aberystwyth*

# Preface to the third edition

We have made further refinements in our goal of offering an elementary textbook with simple instructions in plant tissue culture techniques. Since we are particularly interested in helping plant scientists in developing countries to teach laboratory technicians, our experiments require only basic laboratory equipment and supplies. In this third edition we have subdivided the material into a somewhat more logical arrangement: background information, callus and callus-derived systems, culture organs and organized systems, isolated cells, and supplementary topics. Some innovations include the production of potato callus, the culture of embryos, xylogenesis with isolated mesophyll cells, and a transformation experiment involving hairy-root cultures.

Much appreciated were the services of Dr. Maria Murphy, Editor, Biological Sciences, Cambridge University Press; and Dr. Robin Smith, North American Branch, Cambridge University Press. Our thanks to Jean Norris for supplying certain technical information. We wish to acknowledge and thank the following scientists for their helpful comments: Jeanette Blake, David Burger, Diane Church, Elizabeth Earle, Dave Evans, Hector Flores, Hiroo Fukuda, Esra Galun, Steve Hagen, Raymond Miller, Henry Owen, M. J. R. Rhodes, Carol Stiff, Brent Tisserat, Kazuo Watanabe, and Lindsay Withers. Also, special thanks to Linda Skiba for her help in preparation of the manuscript.

# Preface to the second edition

The need for a new edition of a scientific book is directly related to the influx of new ideas and developments into the field. Solid gains have been made in the past few years in the applications of plant tissue culture techniques to agriculture and industry, and we would like to share some of this information with our readers. Another obvious reason for publishing a new edition is to rectify any errors and misinformation that may have silently crept into the original text.

Although a few organizational changes have been made, the general format of the book remains the same. We have complied with requests to enlarge the section on terminology and definitions. Additional information has been provided on equipment and facilities in Chapter 2. An introduction to the techniques used for the preservation of germplasm will be found in Chapter 15. Since it is hoped that this book will give the reader a broad introduction to the field, we have included a brief nonexperimental chapter on two special topics: virus eradication, and plant tumors and genetic engineering. A table at the end of the book gives the formulations of some tissue culture media: Murashige and Skoog's, Gamborg's B5, White's, and Schenk and Hildebrandt's.

The manuscript of the book was completed by LWR during an appointment as a Senior Visiting Researcher to Queen Elizabeth College, University of London. He gratefully acknowledges the cooperation of Professor P. B. Gahan, Head of the Biology Department.

We are grateful to all reviewers, colleagues, and students who have assisted us by their constructive criticism and suggestions.

# Preface to the first edition

The purpose of this book is to introduce a basic experimental method for each of the major areas of investigation involving the isolation and culture of plant cells, tissues, and organs. Each chapter is devoted to a separate aspect of plant tissue culture, and the chapters are arranged, in general, in order of increasing technical complexity. Although the book was written mainly for use by college undergraduates, research workers from various botanical fields and biology students in high school will also find the text within their grasp. In view of the diverse laboratory facilities that may be available, the experiments selected require a minimum of special equipment. A list of suppliers is given at the end of the book. The book is designed as a laboratory textbook for a course on plant tissue culture techniques, although it may also be used as a supplementary text for developmental botany and biology courses.

The opening chapters present a brief historical survey of the field of plant tissue culture, and a background in sterilization and aseptic techniques. The third chapter examines the various components of the nutrient medium, including inorganic salts, vitamins and other organic supplements, carbohydrates, plant growth regulators, media matrices, and instructions for the preparation of a typical nutrient medium.

The remainder of the text involves laboratory experiments using special culture procedures, and each chapter follows the same format: purpose of the experiment and background information, list of materials, procedure, results, questions for discussion, and selected references. Most of the chapters have an appendix giving additional experiments and techniques. The opening experimental chapters describe the initiation and maintenance of a callus culture, and the preparation of a suspension or liquid culture. Students will repeat White's classic experiment involving the unlimited growth of isolated tomato roots. Another chapter outlines methods for the induction of tracheary element differentiation in cultured tissues. Several chapters introduce diverse approaches to plant propagation by in vitro techniques. The student will attempt to regenerate *Coleus* plants from isolated leaf disks following the cultural principles of organogenesis. The development of embryoids is observed in a suspension culture of carrot cells, and this procedure demonstrates another approach to plant propagation from somatic tissue. A method for the clonal reproduction of plants by the culture of the shoot apex is provided. Isolated protoplasts are prepared and directions given for the induction of protoplast fusion with the creation of somatic hybrids. An important tool in plant breeding involves anther culture, with the resulting production of monoploid plantlets bearing a single set of chromosomes. A brief chapter introduces the concept of using tissue cultures for the commercial production of chemical and medicinal compounds. Finally, some quantitative methods of expressing the results of in vitro experiments are given.

The authors wish to acknowledge the valuable comments received during the preparation of the manuscript, and we particularly thank M. Davey, N. P. Everett, T. Ford, W. P. Hackett, G. G. Henshaw, J. Heslop-Harrison, M. G. K. Jones, A. Komamine, J. O'Hara, V. Raghavan, T. A. Thorpe, J. G. Torrey, G. Wilson, L. A. Withers, and M. M. Yeoman for critical review of the contents. We also wish to thank Miss H. Bigwood and Mr. A. Pugh for their artistic and photographic assistance. The generosity of colleagues who have provided us with original plates and negatives for illustration is acknowledged with the figures.

Most of LWR's contributions to this book were written during an appointment as a Visiting Fellow and Fulbright Senior Scholar to the Australian National University, Canberra. He gratefully acknowledges the excellent cooperation of the Botany Department, the Australian National University, the Research School of Biological Sciences, and the Australian–American Educational Foundation. The authors are also indebted to Florence Roberts for her editorial suggestions.

# PART I

# *Background information*

# 1

# Culture of plant cells, tissues, and organs

*Early attempts, 1902–1939*

The concept that the individual cells of an organism are totipotent is implicit in the statement of the cell theory. Schwann (1839) expressed the view that each living cell of a multicellular organism should be capable of independent development if provided with the proper external conditions (White, 1954). A totipotent cell is one that is capable of developing by regeneration into a whole organism, and this term was probably coined by Morgan in 1901 (Krikorian & Berquam, 1969). The basic problem of cell culture was clearly stated by White (1954): If all the cells of a given organism are essentially identical and totipotent, then the cellular differences observed within an organism must arise from responses of those cells to their microenvironment and to other cells within the organism. It should be possible to restore suppressed functions by isolating the cells from those organismal influences responsible for their suppression. If there has been a loss of certain functions so that the cells in the intact organism are no longer totipotent, then isolation would have no effect on restoring the lost activities. The use of culture techniques enables the scientist to segregate cells, tissues, and organs from the parent organism for subsequent study as isolated biological units. The attempts to reduce an organism to its constituent cells, and subsequently to study these cultured cells as elementary organisms, is therefore of fundamental importance (White, 1954).

Several plant scientists performed experiments on fragments of tissue isolated from higher plants during the latter part of the 19th century. Wound callus formed on isolated stem fragments and root slices was described (Trécul, 1853; Vöchting, 1878; Rechinger, 1893). "Callus" refers to a disorganized proliferated mass of actively dividing cells. Rechinger (1893) examined the "minimum limits" of divisibility of isolated fragments of buds, roots, and other plant material. Although no nutrients were used in these experiments, he concluded that pieces thicker than 1.5 mm were capable of further growth on sand moistened with water. Since isolated fragments thinner that 1.5 mm were apparently incapable of further development, he concluded that this was the size limit beneath which the tissue lost the capability of proliferation. Rechinger reported that the presence of vessel elements appeared to stimulate growth of the fragments. Unfortunately, he did not pursue this clue, since his observations suggested the ability of cambial tissue to proliferate was associated with vascular tissues (Gautheret, 1945).

Haberlandt (1902) originated the concept of cell culture and was the first to attempt to cultivate isolated plant cells in vitro on an artificial medium. A tribute to Haberlandt's genius, with a translation of his paper "Experiments on the culture of isolated plant cells," has been published (Krikorian & Berquam, 1969). Unlike Rechinger, Haberlandt believed that unlimited fragmentation would not influence cellular proliferation. The culture medium consisted mainly of Knop's solution, asparagine, peptone, and sucrose. Although the cultured cells survived for several months, they were incapable of proliferation. Haberlandt's failure to obtain cell division in his cultures was, in part, due to the relatively simple nutrients and to his use of highly differentiated cells. Since Haberlandt did not use sterile techniques, it is difficult to evaluate his results because of the possible effects of bacterial contamination (Krikorian & Berquam, 1969). As example of his genius, Haberlandt suggested the utilization of embryo sac fluids and the possibility of culturing artificial embryos from vegetative cells. In addition, he anticipated the paper-raft technique (Muir, 1953). Following his lack of success with cell cultures, Haberlandt became interested in wound healing. Experiments in this area led to the formulation of his theory of division hormones. Cell division was postulated as being regulated by two hormones: One was "leptohormone," which was associated with vascular tissue, particularly the phloem. The other was a wound hormone released by the injured cells. Subsequent research investigators (Camus, 1949; Jablonski & Skoog, 1954; Wetmore & Sorokin, 1955) verified the association of hormones with vascular tissues.

Early in the twentieth century interest shifted to the culture of meristematic tissues in the form of isolated root tips. These represented the first aseptic organ cultures. Robbins (1922a,b) was the first to develop a technique for the culture of isolated roots and Kotte (1922a,b), a student of Haberlandt's, published similar studies independently. These cultures were of limited success. Robbins and Maneval (1923), with the aid of subcultures, maintained maize roots for 20 weeks. White (1934), experimenting with tomato roots, succeeded for the first time in demonstrating the potentially indefinite culture of isolated roots. According to White (1951), two difficulties hampered the development between 1902 and 1934 of a successful method for culturing excised plant material:

1. the problem of choosing the right plant material, and
2. the formulation of a satisfactory nutrient medium.

With the introduction of root tips as a satisfactory experimental material, the crucial problem became largely one of organic nutrition. White's early success with tomato roots can be attributed to his discovery of the importance of B vitamins, plus the fact that indefinite growth was achieved without the addition of any cell-division factor to the liquid medium.

It is important at this point to make a distinction between organ culture and tissue culture. In the case of excised roots as an example of organ culture, the cultured plant material maintains its morphological identity as a root with the same basic anatomy and physiology as in the roots of the parent plant. There are some exceptions, and slight changes in anatomy and physiology may occur during the culture period. According to Street (1977a), the term "tissue culture" can be applied to any multicellular culture growing on a solid medium (or attached to a substratum and nurtured with a liquid medium) that consists of many cells in protoplasmic continuity. Typically, the culture of an explant, consisting of one or more tissues, results in a callus that has no structural or functional counterpart with any tissue of the normal plant body.

The first plant tissue cultures, in the sense of long-term cultures of callus, involved explants of cambial tissues isolated from carrot (Gautheret, 1939; Nobécourt, 1939) and tobacco tumor tissue from the hybrid *Nicotiana glauca* × *N. langsdorffii* (White, 1939). The latter tumor tissue requires no exogenous cell-division factor. Results from these three laboratories, published independently, appeared almost simultaneously. Fortunately, plant physiologists working in other areas had discovered some of the hormonal characteristics of indole-3-acetic acid, IAA (Snow, 1935; Went & Thimann, 1937), and the addition of this auxin to the culture medium was essential to the success of the carrot cultures maintained by Nobécourt and Gautheret. According to Gautheret (1939), the carrot cultures required Knop's solution supplemented with Bertholot's salt mixture, glucose, gelatine, thiamine, cysteine-HCl, and IAA (White, 1941). The goal at that time was to demonstrate the potentially unlimited growth of a given culture, by repeated subcultures, with the formation of undifferentiated callus. The workers were fascinated by the apparent immortality of their cultures and devoted much effort to determining the nutritional requirements for sustained growth.

*Basic studies on nutrition and morphogenesis, 1940–1978*

Because of the lull in botanical research during the war years (1939–45), relatively little was accomplished until a resurgence of interest in the early 1950s. Some fundamental studies, however, were undertaken. White and Braun (1942) initiated experiments on crown gall and tumor formation in plants. Probably the most significant event leading to advancement in the next decade was the discovery of the nutritional quality of liquid endosperm extracted from coconut. Following the success of Van Overbeek, Conklin, and Blakeslee (1941) with the culture of isolated *Datura* embryos on a medium enriched with coconut milk, other workers rapidly adopted this natural plant extract. The combination of coconut milk and 2,4-D had a remarkable effect on the proliferation of cultured carrot and potato tissues (Caplin & Steward, 1948; Steward & Caplin, 1951, 1952). Although it was first thought that a single substance, termed the "coconut-milk factor," was involved as a growth stimulant, several constituents were later found responsible for its activity. Steward's group at Cornell University made numerous contributions in technique, nutrition, quantitative analyses of culture growth, and morphogenesis. The regeneration of carrot plantlets from cultured secondary-phloem cells of the taproot clearly demonstrated the totipotency of plant cells (Krikorian, 1975). The phenomenon of somatic embryogenesis in carrot cultures was discovered at approximately the same time by Steward (1958) and Reinert (1959). The consequences of this discovery is discussed in Chapter 8.

The discovery of cytokinins stems from Skoog's tissue culture investigations at the University of Wisconsin. During attempts to induce unlimited callus production from mature tobacco pith cells, numerous compounds were tested for possible activity in stimulating cell division. Although coconut water or yeast extract plus IAA promoted cell division, efforts were made to locate a specific cell-division factor. Since adenine, in the presence of auxin, was found to be active in stimulating callus growth and bud formation in tobacco cultures (Skoog & Tsui, 1948; Sterling, 1950), nucleic acids were then examined. Skoog's group eventually located a potent cell-division factor in degraded DNA preparations. It was isolated, identified as 6-furfurylaminopurine, and named "kinetin" (Miller et al., 1955). The related analogue, 6-benzylaminopurine, was then synthesized, and it too stimulated cell division in cultured tissues. The generic term "cytokinin" was given to this group of 6-substituted aminopurine compounds that stimulate cell division in cultured plant tissues and behave in a physiological manner similar to kinetin. Later it was discovered that zeatin, isopentyl adenine, and other cytokinins are naturally occurring plant hormones. Often these compounds are attached to ribose sugar ("ribosides") or to ribose and phosphate ("ribotides"). The stimulatory properties of coconut water are partly due to the presence of zeatin riboside. Skoog and Miller (1957) advanced the hypothesis that shoot and root initiation in cultured callus can be regulated by varying the ratio of auxin and cytokinin in the medium (Chapter 6). In addition to the cytokinins, other endogenous cell-division factors may exist in plant tissues (Wood et al., 1969).

It was found that callus fragments, transferred to a liquid medium and aerated on a shaker, gave a suspension of single cells and cell aggregates that could be propagated by subculture (Muir, 1953; Muir, Hildebrandt, & Riker, 1954). Steward's group made extensive use of carrot suspension cultures, and it became evident that this technique offered much potential for studying many facets of cell biology and biochemistry (Nickell, 1956). Street and coworkers have pioneered the development of various procedures for the culture of cell suspensions (e.g., chemostats and turbidostats; Chapter 7). Torrey and his colleagues also conducted studies on cell suspensions (Torrey & Shigomura, 1957; Torrey & Reinert, 1961; Torrey, Reinert, & Merkel, 1962).

Muir (1953) succeeded in developing a technique for the culture of single isolated cells. Single cells were placed on squares of filter paper, and the lower surface of the paper was placed in contact with an actively growing "nurse" culture. This paper-raft nurse technique provided the isolated cells not only with nutrients from the medium via the older culture, but also with growth factors synthesized by the nurse tissue. Although Muir's experiments involved bacteria-free crown gall (*Agrobacterium tumefaciens*) tumor cells, single-cell clones were produced later from normal cells. In another approach, a single cell was suspended as a hanging drop in a microchamber (Torrey, 1957; Jones et al., 1960). The agar-plating method of Bergmann (1960) involved separating a single-cell fraction by filtration, mixing the cells with warm agar, and then plating the cells as a thin layer in a Petri dish. These early investigations with single-cell cultures have been reviewed by Street (1977b) and Hildebrandt (1977). Although Muir and his colleagues reported in 1954 that single isolated cells exhibited cell division, this claim was contested by De Ropp (1955). Subsequent investigations provided the necessary evidence that single cells are capable of proliferation (Torrey, 1957; Muir et al., 1958). The question of totipotency was completely resolved by Vasil and Hildebrandt (1965) by the demonstration that a single isolated cell can divide and ultimately give rise to a whole plant.

Plant tissue cultures have been used extensively for the study of cytodifferentiation, particularly the formation of tracheary elements in cultured tissues. A variety of different techniques have been employed. Wedges of agar containing auxin and sucrose "grafted" to a block of callus induced tracheary element formation (Wetmore & Sorokin, 1955; Wetmore & Rier, 1963; Jeffs & Northcote, 1967). Primary explants from many different plant tissues are capable of producing tracheary elements during culture on agar or in a liquid medium (Chapter 12). The use of plant tissue cultures to study vascular differentiation was reviewed by Roberts (1988).

Many of the early investigators employed either herbaceous or woody dicot tissues as sources of primary explants, although other groups of plants were also used as experimental material. Morel

(1950) successfully cultured monocot tissues with the aid of coconut water. Ball (1955) cultured tissues of the gymnosperm *Sequoia sempervirens*, and Tulecke prepared haploid cultures from the pollen of *Taxus* (1959) and *Ginkgo biloba* (1953, 1957). Harvey and Grasham (1969) published media requirements for establishing cultures of 12 conifer species. Tissue culture procedures have been used in developmental anatomy studies involving excised shoot apices of lower plants (e.g., ferns, *Selaginella*, and *Equisetum;* Wetmore & Wardlaw, 1951).

During the 1960s it was shown that cultured pollen and the microsporogenous tissue of anthers have the potential to produce vast numbers of haploid embryos (Guha & Maheshwari, 1966, 1967; Bourgin & Nitsch, 1967). Later, with a technique developed by C. Nitsch, it became possible to culture microspores of *Nicotiana* and *Datura*, to double the chromosome number of the microspores, and to collect seeds from the homozygous diploid plants within a five-month period (Nitsch, 1974, 1977). Although there are technical problems associated with this technique, haploid cultures have been used successfully in China for the selection of improved varieties of crop plants (Chapter 11).

Another important development during the 1960s was the enzymatic isolation and culture of protoplasts (Cocking, 1960). This method involves removing the cell wall with purified preparations of cellulase and pectinase, while regulating protoplast expansion with an external osmoticum. The cultured protoplasts regenerate new cell walls, form cell colonies, and ultimately form plantlets (Takebe, Labib, & Melchers, 1971). Some of the experimental approaches currently employed include

1. protoplast fusion within species, between species, between monocots and dicots, and even between plants and animals;
2. introduction of mitochondria and plastids into protoplasts;
3. uptake of blue-green algae, bacteria, and viruses by protoplasts; and
4. the transfer of genetic information into isolated protoplasts.

Cells containing foreign organelles are termed "cytoplasmic hybrids" or "heteroplasts," whereas cells containing transferred nuclei are referred to as "heterokaryons" or "heterokaryocytes." The fusion of the two nuclei in heterokaryons produces hybrid cells (Chapter 14).

Plant tissue culture techniques have been widely used for the commercial propagation of plants. Most of the applications are based on the characteristic of cytokinins to stimulate bud proliferation in the cultured shoot apex. Ball (1946) demonstrated the possibility of regenerating plants from isolated explants of angiosperm shoot apices. Later, Wetmore and Morel regenerated whole plants from shoot apices measuring 100–250 $\mu$m in length and bearing one or two leaf primordia (Wetmore & Wardlaw, 1951). Modifications of Morel's (1960) shoot-apex technique have been used for orchid propagation (Morel, 1964). Premixed culture media, specifically formulated for the propagation of certain plants, are commercially available (see appended Commercial Sources of Supplies). Several recent publications are devoted to technical problems associated with the mass propagation of higher plants (Vasil, 1980; Conger, 1981; Constantin et al., 1981; Thorpe, 1981; Bonga & Durzan, 1982; Tomes et al., 1982). This subject is discussed further in Chapter 10.

One of the earliest applications of plant tissue culture involved the study of plant tumor physiology. White and Braun (1942) reported the growth of bacteria-free crown gall tissue. Braun has devoted his career at Rockefeller University to the study of plant cancer (Braun, 1974, 1975), and a summary of his research has appeared (Butcher, 1977). Related to these studies is the phenomenon of habituation. Gautheret (1946) observed that a callus culture of *Scorzonera hispanica*, which originally required auxin in the medium for growth, often developed outgrowths of callus that would grow indefinitely on an auxin-deficient medium. The term "habituation" refers to inherited changes in nutritional requirements arising in cultured cells, especially changes involving plant hormones. For example, an auxin-habituated culture has lost its original requirement for exogenous auxin (Butcher, 1977). These cultures are important in investigations of plant cancer. The grafting of auxin- and cytokinin-habituated tissues into healthy plants produces tumors (Butcher, 1977). Tissue culture techniques have also been used to produce pathogen-free plants via apical meristem cultures. Ingram and Helgeson (1980) reviewed the applications of in vitro procedures to plant pathology.

An interesting account of the early development in both plant and animal cell culture is given in White's book (1954), and Gautheret's monumental work (1959) was an invaluable guide to the early investigators. A summary of the pioneer studies on plant morphogenesis involving cultured tissues was compiled by Butenko (1964), and Gautheret (1983) has recalled some of the highlights of the field.

### Emergence of a new technology, circa 1978 to the present

By the late 1970s it had become evident that plant tissue culture technology was beginning to make significant contributions to agriculture and industry (Murashige, 1978; Zenk, 1978). In agriculture, the major areas are haploid breeding, clonal propagation, mutant cultures, pathogen-free plants, production of secondary products, and genetic engineering. In addition, the cryopreservation of plant tissue cultures and the establishment of in vitro gene banks have attained considerable interest (Chapter 15).

The greatest success has been achieved with in vitro clonal propagation. In vitro techniques have revitalized the orchid industry. Murashige (1977) estimated that over six hundred species of ornamental plants have been cloned. Cloning has been extended to forest trees, fruit trees, oil-bearing plants, vegetables, and numerous agronomic crop plants. Clonal propagation of potato plants has been achieved on a large scale by regeneration from isolated leaf-cell protoplasts (Shepard, 1982). By 1982 more than a hundred tissue culture facilities were engaged in the commercial propagation of plants (Loo, 1982). In other areas, plant tissue culture has had only limited economic success. Haploid breeding has produced relatively few established cultivars, mainly because of the low frequency of the appearance of new agriculturally important genotypes by this method. Progress is being made, however, particularly in China (Loo, 1982). The application of mutagenic agents to cultures, followed by suitable screening techniques, has led to the regeneration of mutant plants showing disease or stress resistance. Chaleff (1983) reviewed the technical problems associated with this approach. Several pathogen-free plants have been developed (Murashige, 1978), and tissue culture technology now plays an important role in plant pathology. For example, protoplasts are currently employed for the study of virus infection and biochemistry (Rottier, 1978). The prospects of success with the genetic manipulation of plants have created considerable public interest, and advances to date are now considerable. Several approaches have been considered. One technique involves the transfer of genetic information by the fusion of protoplasts isolated from two different organisms. This method provides the opportunity of producing hybrids between related sexually incompatible species. Melchers, Sacristan, and Holder (1978) produced a somatic hybrid plant from the fusion of potato and tomato protoplasts, both members of the Solanaceae family. It appears unlikely, however, that this technique will yield any plants of economic importance in the near future. Another technique concerns the insertion of foreign genes attached to a plasmid vector into the naked protoplast (Barton & Brill, 1983; Dodds & Bengochea, 1983). The transfer of the plasmid DNA into the protoplast is usually accomplished by means of liposomes. A major problem, however, is whether the inserted gene will be integrated into the host genome, transcribed, and expressed in the mature plant. Several research groups have produced transformed tobacco plants following single-cell transformation (i.e., gene insertion) (Wullems et al., 1982; Chilton, 1983). The use of *Agrobacterium*-mediated plant transformation, and the use of the so-called gene gun literally to shoot DNA into plant cells, has led to the development of many novel plants. The first of these was commercialized by Calgene in 1994.

Industrial applications involve large-scale suspension cultures capable of synthesizing significant amounts of useful compounds. These secondary products of industrial interest include antimicrobial compounds, antitumor alkaloids, food flavors, sweeteners, vitamins, insecticides, and enzymes. A major problem has been the genetic instability of the cultures, as well as engineering problems associated with this technique. Nevertheless, progress has been made at the industrial level, especially by the Japanese (Misawa, 1977, 1980). This topic is discussed further in Chapter 16.

SELECTED REFERENCES

Ball, E. (1946). Development in sterile culture of stem tips and subjacent regions of *Tropaeolum majus* L. and *Lupinus albus* L. *Am. J. Bot.* 33, 301–18.

(1955). Studies on the nutrition of the callus culture of *Sequoia senpervirens*. *Ann. Biol. 31*, 281–305.
Barton, K. A., & Brill, W. J. (1983). Prospects in plant genetic engineering. *Science 219*, 671–6.
Bergmann, L. (1960). Growth and division of single cells of higher plants in vitro. *J. Gen. Physiol. 43*, 841–51.
Bonga, J. M., & Durzan, D. J. (eds.) (1982). *Tissue culture in forestry*. The Hague: Martinus Nijhoff/Junk.
Bourgin, J. P., & Nitsch, J. P. (1967). Obtention de *Nicotiana* haploides à oartir d'étamines cultivées in vitro. *Ann. Physiol. Vég. Paris 9*, 377–82.
Braun, A. C. (1974). *The biology of cancer*. New York: Addison–Wesley.
 (1975). The cell cycle and tumorigenesis in plants. In *Cell cycle and cell differentiation*, ed. J. Reinert & H. Holtzer, pp. 177–96. Berlin: Springer–Verlag.
Butcher, D. N. (1977). Plant tumor cells. In *Plant tissue and cell culture*, ed. H. E. Street, pp. 429–61. Oxford: Blackwell Scientific Publications.
Butenko, R. G. (1964). *Plant tissue culture and plant morphogenesis*. Translated from the Russian. Jerusalem: Israel Program for Scientific Translation, 1968.
Camus, G. (1949). Recherches sur le rôle des bourgeons dans les phénomènes de morphogénèse. *Rev. Cytol. Biol. Vég. 9*, 1–199.
Caplin, S. M., & Steward, F. C. (1948). Effect of coconut water on the growth of explants from carrot root. *Science 108*, 655–7.
Chaleff, R. S. (1983). Isolation of agronomically useful mutants from plant cell cultures. *Science 219*, 676–82.
Chilton, M. D. (1983). A vector for introducing new genes into plants. *Sci. Am. 248*, 50–9.
Cocking, E. C. (1960). A method for the isolation of plant protoplasts and vacuoles. *Nature 187*, 962–3.
Conger, B. V. (ed.) (1981). *Cloning agricultural plants via in vitro techniques*. Boca Raton: CRC Press.
Constantin, M. J., Henke, R. R., Hughes, K. W., & Conger, B. V. (eds.) (1981). Propagation of higher plants through tissue culture: Emerging technologies and strategies. *Environ. Exp. Bot. 21*, 269–452.
De Ropp, R. S. (1955). The growth and behavior in vitro of isolated plant cells. *Proc. Roy. Soc. B 144*, 86–93.
Dodds, J. H., & Bengochea, T. (1983). Principles of plant genetic engineering. *Outlook Agric. 12*, 16–20.
Gautheret, R. J. (1939). Sur la possibilité de réaliser la culture indéfinie des tissu de tubercules de carotte. *C. R. Acad. Sci. (Paris) 208*, 118–21.
 (1945). *La culture des tissus*. Paris: Gallimard.
 (1946). Comparison entre l'actions de l'acide indoleacetique et celle du *Phytomonas tumefaciens* sur la croissance des tissus végétaux. *C. R. Soc. Biol. (Paris) 140*, 169–71.
 (1959). *La culture des tissus végétaux: Techniques et réalisations*. Paris: Masson.

(1983). Plant tissue culture: A history. *Bot. Mag. Tokyo 96*, 393–410.
Guha, S., & Maheshwari, S. C. (1966). Cell division and differentiation of embryos in the pollen grains of *Datura* in vitro. *Nature 212*, 97–8.
 (1967). Development of embryoids from pollen grains of *Datura* in vitro. *Phytomorphology 17*, 454–61.
Haberlandt, G. (1902). Kulturversuche mit isolierten Pflanzenzellen. *Sitzungsber. Akad. Wiss. Wien, Math. Nat. Classe 111*, Abt.1, 69–92.
Harvey, A. E., & Grasham, J. L. (1969). Procedure and media for obtaining tissue cultures of twelve conifer species. *Can. J. Bot. 47*, 547–9.
Hildebrandt, A. C. (1977). Single cell culture, protoplasts and plant viruses. In *Applied and fundamental aspects of plant cell, tissue, and organ culture*, ed. J. Reinert & Y. P. S. Bajaj, pp. 581–97. Berlin: Springer-Verlag.
Ingram, D. S., & Helgeson, J. P. (eds.) (1980). *Tissue culture methods for plant pathologists*. Oxford: Blackwell Scientific Publications.
Jablonski, J., & Skoog, F. (1954). Cell enlargement and division in excised tobacco pith tissue. *Physiol. Plant. 7*, 16–24.
Jeffs, R. A., & Northcote, D. H. (1967). The influence of indol-3yl-acetic acid and sugar on the pattern of induced differentiation in plant tissue cultures. *J. Cell Sci. 2*, 77–88.
Jones, L. E., Hildebrandt, A. C., Riker, A. J., & Wu, J. H. (1960). Growth of somatic tobacco cells in microculture. *Am. J. Bot. 47*, 468–75.
Kotte, W. (1922a). Wurzelmeristem in Gewebekultur. *Ber. Dtsch. Bot. Ges. 40*, 269–72.
 (1922b). Kulturversuch isolierten Wurzelspitzen. *Beitr. Allg. Bot. 2*, 413–34.
Krikorian, A. D. (1975). Excerpts from the history of plant physiology and development. In *Historical and current aspects of plant physiology: A symposium honoring F. C. Steward*, ed. P. J. Davies, pp. 9–97. Ithaca, N.Y.: Cornell University Press.
Krikorian, A. D., & Berquam, D. L. (1969). Plant cell and tissue cultures: The role of Haberlandt. *Bot. Rev. 35*, 59–88.
Loo, S. (1982). Perspective on the application of plant cell and tissue culture. In *Plant tissue culture 1982*, ed. A Fujiwara, pp. 19–24. Tokyo: Japanese Association for Plant Tissue Culture.
Melchers, G., Sacristan, M. D., & Holder, A. A. (1978). Somatic hybrid plants of potato and tomato regenerated from fused protoplasts. *Carlsberg Res. Commun. 43*, 203–18.
Miller, C. O., Skoog, F., Saltza, M., & Strong, F. M. (1955). Kinetin, a cell division factor from desoxyribonucleic acid. *J. Am. Chem. Soc. 77*, 1392.
Misawa, M. (1977). Production of natural substances by plant cell cultures described in Japanese patents. In *Plant tissue culture and its bio-technological application*, ed. W. Barz, E. Reinhard, & M. H. Zenk, pp. 17–26. Berlin: Springer–Verlag.

(1980). Industrial and government research. In *Plant tissue culture as a source of biochemicals*, ed. E. J. Staba, pp. 167–90. Boca Raton: CRC Press.
Morel, G. (1950). Sur la culture des tissus de deux Monocotylédones. *C. R. Acad. Sci. (Paris) 230*, 1099–101.
  (1960). Producing virus-free Cymbidiums. *Am. Orchid Soc. Bull. 29*, 495–7.
  (1964). Tissue culture – a new means of clonal propagation in orchids. *Am. Orchid Soc. Bull. 33*, 473–8.
Morgan, T. H. (1901). *Regeneration*. London: Macmillan.
Muir, W. H. (1953). Culture conditions favoring the isolation and growth of single cells from higher plants in vitro. Ph.D. thesis, Dept. of Plant Pathology, University of Wisconsin, Madison.
Muir, W. H., Hildebrandt, A. C., & Riker, A. J. (1954). Plant tissue cultures produced from single isolated plant cells. *Science 119*, 877–87.
  (1958). The preparation, isolation and growth in culture of single cells from higher plants. *Am. J. Bot. 45*, 589–97.
Murashige, T. (1977). Clonal crops through tissue culture. In *Plant tissue culture and its bio-technological application*, ed. W. Barz, E. Reinhard, & M. H. Zenk, pp. 392–403. Berlin: Springer–Verlag.
  (1978). The impact of plant tissue culture on agriculture. In *Frontiers of plant tissue culture 1978*, ed. T. A. Thorpe, pp. 15–26. Calgary: IAPTC.
Nickell, L. G. (1956). The continuous submerged cultivation of plant tissues as single cells. *Proc. Natl. Acad. Sci. USA 42*, 848–50.
Nitsch, C. (1974). La culture de pollen isolé sur milieu synthétique. *C. R. Acad. Sci. (Paris) 278*, 1031–4.
  (1977). Culture of isolated microspores. In *Applied and fundamental aspects of plant cell, tissue, and organ culture*, ed. J. Reinert & Y. P. S. Bajaj, pp. 268–78. Berlin: Springer–Verlag.
Nobécourt, P. (1939). Sur la perennite et l'augmentation de volume des cultures de tissus végétaux. *C. R. Soc. Biol. (Paris) 130*, 1270–1.
Rechinger, C. (1893). Untersuchungen über die Grenzen der Teilbarkeit im Pflanzenreich. *Abh. Zool.-Bot. Ges. (Vienna) 43*, 310–34.
Reinert, J. (1959). Über die Kontrolle der Morphogenese und die Induktion von Adventiveembryonen an Gewebekulturen aus Karotten. *Planta 53*, 318–33.
Robbins, W. J. (1922a). Cultivation of excised root tips and stem tips under sterile conditions. *Bot. Gaz. 73*, 376–90.
  (1922b). Effect of autolysed-yeast and peptone on growth of excised corn root tips in the dark. *Bot. Gaz. 74*, 59–79.
Robbins, W. J., & Maneval, W. E. (1923). Further experiments on growth of excised root tips under sterile conditions. *Bot. Gaz. 76*, 274–87.
Roberts, L. W. (1988). Evidence from wound responses and tissue cultures. In *Vascular differentiation and plant growth regulators*, ed. L. W. Roberts, P. B. Gahan, & R. Aloni. Heidelberg: Springer–Verlag, pp. 63–88.

Rottier, J. M. (1978). The biochemistry of virus multiplication in leaf cell protoplasts. In *Frontiers of plant tissue culture 1978*, ed. T. A. Thorpe, pp. 255–64. Calgary: IAPTC.

Schwann, Th. (1839). *Mikroskepische Untersuchungen über die Übereinstimmung in der struktur und dem Waschstume der Tiere und Pflanzen.* Leipzig: W. Englemann, No. 176, Oswalds Klassiker der exakten Wissenschaften, 1910.

Shepard, J. F. (1982). The regeneration of potato plants from leaf-cell protoplasts. *Sci. Am. 246,* 154–66.

Skoog, F., & Miller, C. O. (1957). Chemical regulation of growth and organ formation in plant tissues cultured in vitro. *Symp. Soc. Exp. Biol. 11,* 118–30.

Skoog, F., & Tsui, C. (1948). Chemical regulation of growth and organ formation in plant tissues cultured in vitro. *Am. J. Bot. 35,* 782–7.

Snow, R. (1935). Activation of cambial growth by pure hormones. *New Phytol. 34,* 347–59.

Sterling, C. (1950). Histogenesis in tobacco stem segments cultured *in vitro*. *Am. J. Bot. 37,* 464–70.

Steward, F. C. (1958). Growth and development of cultivated cells. III. Interpretations of the growth from free cell to carrot plant. *Am. J. Bot. 45,* 709–13.

Steward, F. C., & Caplin, S. M. (1951). A tissue culture from potato tubers. The synergistic action of 2,4-D and coconut milk. *Science 113,* 518–20.

  (1952). Investigation on growth and metabolism of plant cells: IV. Evidence on the role of the coconut-milk factor in development. *Ann. Bot. (Lond.) 16,* 491–504.

Street, H. E. (1977a). Introduction. In *Plant tissue and cell culture*, ed. H. E. Street, pp. 1–10. Oxford: Blackwell Scientific Publications.

  (1977b). Single-cell clones – derivation and selection. In *Plant tissue and cell culture*, ed. H. E. Street, pp. 207–22. Oxford: Blackwell Scientific Publications.

Takebe, I., Labib, G., & Melchers, G. (1971). Regeneration of whole plants from isolated mesophyll protoplasts of tobacco. *Naturwissenschaften 58,* 318–20.

Thorpe, T. A. (ed). (1981). *Plant tissue culture: Methods and applications in agriculture.* New York: Academic Press.

Tomes, D. T., Ellis, B. E., Harney, P. M., Kasha, K. J., & Peterson, R. L. (eds). (1982). *Application of plant cell and tissue culture to agriculture and industry.* Guelph, Ontario: University of Guelph Press.

Torrey, J. G. (1957). Cell division in isolated single cells in vitro. *Proc. Natl. Acad. Sci. USA 43,* 887–91.

Torrey, J. G., & Reinert, J. (1961). Suspension cultures of higher plant cells in synthetic medium. *Plant Physiol. 36,* 483–91.

Torrey, J. [G.], Reinert, J., & Merkel, N. (1962). Mitosis in suspension cultures of higher plant cells in synthetic medium. *Am. J. Bot. 49,* 420–5.

Torrey, J. G., & Shigomura, J. (1957). Growth and controlled morphogenesis in pea root callus tissue grown in liquid media. *Am. J. Bot.* 44, 334–44.

Trécul, M. (1853). Accroissement des végétaux dicotylédones ligneux, reproduction du bois et de l'écorce par le bois décortiqué. *Ann. Sci. Nat. Bot.*, ser. III, 19, 157–92.

Tulecke, W. (1953). A tissue derived from the pollen of *Ginkgo biloba*. *Science* 117, 599–600.

(1957). The pollen of *Ginkgo biloba*: In vitro culture and tissue formation. *Am. J. Bot.* 44, 602–8.

(1959). The pollen cultures of C. D. La Rue: A tissue from pollen of *Taxus*. *Bull. Torrey Bot. Club* 86, 283–9.

Van Overbeek, J., Conklin, M. E., & Blakeslee, A. F. (1941). Factors in coconut milk essential for growth and development of very young *Datura* embryos. *Science* 94, 350–1.

Vasil, I. K. (ed). (1980). Perspectives in plant cell and tissue culture. *Int. Rev. Cytol. Suppl.* 11A–B (2 vols.). New York: Academic Press.

Vasil, V., & Hildebrandt, A. C. (1965). Differentiation of tobacco plants from single, isolated cells in microculture. *Science* 150, 889–92.

Vöchting, H. (1878). *Über Organbildung im Pflanzenreich*. Bonn: Verlag von Max Cohen & Sohn.

Went, F. W., & Thimann, K. V. (1937). *Phytohormones*. New York: Macmillan.

Wetmore, R. H., & Rier, J. P. (1963). Experimental induction of vascular tissues in callus of angiosperms. *Am. J. Bot.* 50, 418–30.

Wetmore, R. H., & Sorokin, S. (1955). On the differentiation of xylem. *J. Arnold Arbor. Harv. Univ.* 36, 305–24.

Wetmore, R. H., & Wardlaw, C. W. (1951). Experimental morphogenesis in vascular plants. *Annu. Rev. Plant Physiol.* 2, 269–92.

White, P. R. (1934). Potentially unlimited growth of excised tomato root tips in a liquid medium. *Plant Physiol.* 9, 585–600.

(1939). Potentially unlimited growth of excised plant callus in an artificial medium. *Am. J. Bot.* 26, 59–64.

(1941). Plant tissue culture. *Biol. Rev.* 16, 34–48.

(1951). Nutritional requirements of isolated plant tissues and organs. *Annu. Rev. Plant Physiol.* 2, 231–44.

(1954). *The cultivation of animal and plant cells*. New York: Ronald Press (2d Ed., 1963).

White, P. R., & Braun, A. C. (1942). A cancerous neoplasm of plants. Autonomous bacteria-free crown-gall tissue. *Cancer Res.* 2, 597–617.

Wood, H. N., Braun, A. C., Brandes, H., & Kende, H. (1969). Studies on the distribution and properties of a new class of cell division-prompting substances from higher plant species. *Proc. Natl. Acad. Sci. USA* 62, 349–56.

Wullems, G., Krens, F., Peerbolte, R., & Shilperoort, R. (1982). Transformed tobacco plants regenerated after single cell transformation. In *Plant tis-*

*sue culture 1982*, ed. A. Fujiwara, pp. 505–6. Tokyo: Japanese Association for Plant Tissue Culture.

Zenk, M. H. (1978). The impact of plant tissue culture on industry. In *Frontiers of plant tissue culture 1978*, ed. T. A. Thorpe, pp. 1–13. Calgary: IAPTC.

# 2

## Laboratory facilities

A laboratory devoted to in vitro procedures with plant tissues must have adequate space for the performance of several functions. According to White (1963), it must provide facilities for

1. media preparation, sterilization, cleaning, and storage of supplies;
2. aseptic manipulation of plant material;
3. growth of the cultures under controlled environmental conditions;
4. examination and evaluation of the cultures; and
5. assembling and filing of records.

The grouping of functions will vary considerably from one laboratory to another. The ideal organization will allow a separate room for each of the following functions (White, 1963): media preparation, aseptic procedures, incubation of cultures, and general laboratory operations. If one has the opportunity to plan an in vitro laboratory in advance, the component facilities should be arranged as a production line (Street, 1973). The area involved with washing and storage of glassware should lead to the facilities for oven sterilization and media preparation. Materials should then move from autoclave sterilization to the aseptic transfer facility. After the aseptic operations, the cultures are transferred to incubators or controlled-environment chambers. The cultures should be in close proximity to the laboratory containing microscopes and facilities for evaluation of the results. Discarded and contaminated cultures are transferred back to the washing area. It is of utmost importance to give careful consideration to the arrangement of the aseptic procedures. Because most laboratories do not have a separate sterile room, some type of laminar flow cabinet or bacteriological glove box is required. A novel approach to this is the Cleansphere shown in Figure 2.1 This facility must be located in an area free from drafts and with a minimum of traffic. In addition, it should not be located in the vicinity of scientists from other research groups working with airborne microorganisms (Chapter 3).

*Dishwashing.* Discarded cultures, as well as contaminated ones, are autoclaved briefly in order to liquefy the agar and to kill any contaminants that may be present. The culture glassware is easier to wash after the spent medium has been liquefied and removed.

Figure 2.1. An alternative to a laminar flow hood is a glove box such as the Cleansphere®. (Photo courtesy of National Labnet Company.)

After scrubbing with a brush in a hot detergent bath, the glassware is rinsed repeatedly with tap water, and then given two or three rinses in distilled water (de Fossard, 1976). If an automatic dishwasher machine is used, a final rinse with distilled or demineralized water should be used to remove any possible traces of detergent. After washing, the glassware is oven dried prior to storage. Certain cell cultures require scrupulously clean culture vessels; therefore, a routine dishwashing program is inadequate. New glassware may release chemicals that are toxic to the cultured tissues. Additional information can be obtained from Street (1973) and Biondi and Thorpe (1981).

*Media preparation.* Although media preparation requires a balance sensitive to milligram quantities for weighing hormones and vitamins, a less sensitive scale may be used for weighing agar and carbohydrates. The media reagents should be shelved near the balance for convenience. A refrigerator and a freezer in the media room is necessary for storing stock solutions and chemicals that degrade at room temperature. A combination hot plate and magnetic stirrer is a time saver for dissolving inorganic reagents. Either a pH meter or pH indicator paper is required for adjusting the final pH of the medium. Relatively large quantities of single- and double-distilled water must be available in the media room.

Sterilization equipment is an integral part of media preparation. A commercial electric stove is the most economical type of oven sterilization. Wet-heat sterilization involves either an autoclave or a pressure cooker. Some hormones and vitamins are sterilized by ultrafiltration at room temperature. After sterilization of the culture vessels by dry heat and autoclaving the medium, the culture tubes or Petri dishes are poured in the transfer chamber. It is also useful to have a microwave oven to dissolve agar and melt medium. The subsequent aseptic techniques are discussed in Chapters 3 and 4.

*Incubation of the cultures.* The freshly prepared cultures are grown under carefully regulated environmental conditions; that is, temperature, light, and humidity. This is accomplished with an incubator, plant growth chamber, or controlled environment room (Fig. 2.2). If cell suspensions are cultured, some type of shaker (Fig. 2.3) or aeration equipment will be necessary. Several engineering aspects should be considered in designing a culture room: safety and convenience of the electrical system, air flow for uniform temperature regulation, arrangement of the shelving, and elimination of airborne contaminants (Wetherell, 1982). Optimal environmental conditions will vary depending on the species and the purpose of the experiment, and consideration should be given to diurnal temperature variations, light intensity, light quality, and photoperiod (i.e., relative length of light–dark cycles). Fluorescent lamps have certain advantages over incandescent sources: a better spectral quality, a more convenient shape, and a lower heat output. When constructing a growth room, it is good to keep the light bulbs outside, as they generate a lot of heat. Some cultures, however, appear to show the best growth in the presence of a mixture of both types of illumination. Experiments conducted by Murashige (1974) with *Asparagus, Gerbera, Saxifraga,* and bromeliads indicated an optimum light intensity of 1,000 lux during culture initiation and shoot proliferation. A higher optimum of 3,000–10,000 lux was required for the establishment of plantlets. These experiments utilized Gro-Lux or white fluorescent lamps with a daily exposure period of 16 hr (Murashige, 1974). It is advisable to equip the culture room with a clock-operated timing switch for the regulation of photoperiods. Although some investigators may want to expose the cultures to thermoperiodic cycles, most experiments are conducted with constant temperatures set

Figure 2.2. A walk-in controlled environment room for incubation of in vitro cultures. (International Potato Center.)

at approximately 25–27 °C. Some morphogenetic responses are evidently sensitive to temperature fluctuations (Murashige, 1974). Additional information on the effects of various light sources on plant growth can be found in the review by Cathey and Campbell (1982).

*Additional needs.* General laboratory requirements vary considerably, depending on the type of data required. A suitable hand lens and dissection microscope are important for the macroexamination of the cultures, and a compound microscope equipped with photomicrograph accessories is available in most laboratories. A chemical hood should be used with maceration procedures involving chromic acid, as well as for the storage of volatile and potentially dangerous chemicals. Protoplast purification requires a bench-top centrifuge. Also, an inverted microscope is helpful. All plant tissue culture laboratories should be equipped with a fire extinguisher and a first aid kit.

Descriptions of the facilities used by the early investigators can be found in the publications of Gautheret (1959), White (1963),

Figure 2.3. Orbital shaker for the aeration of liquid cultures. (Courtesy of Lab-Line Instruments.)

and Butenko (1964). Street (1973), de Fossard (1976), Biondi and Thorpe (1981), Bonga (1982), and Wetherell (1982) describe modern laboratory facilities.

Possibly the best way of deciding on the type of equipment and facilities for plant tissue culture is to arrange a visit to a plant tissue culture laboratory (de Fossard, 1976). Addresses of laboratories may be obtained from the national correspondent of the country's branch of the International Association for Plant Tissue Culture (IAPTC). The name and address of the national correspondent can be obtained from the IAPTC secretary.

SELECTED REFERENCES

Biondi, S., & Thorpe, T. A. (1981). Requirements for a tissue culture facility. In *Plant tissue culture. Methods and applications in agriculture*, ed. T. A. Thorpe, pp. 1–20. New York: Academic Press.

Bonga, J. M. (1982). Tissue culture techniques. In *Tissue culture in forestry*, ed. J. M. Bonga & D. J. Durzan, pp. 4–35. The Hague: Martinus Nijhoff/Junk.

Butenko, R. G. (1964). *Plant tissue culture and plant morphogenesis*. Translated from the Russian. Jerusalem: Israel Program for Scientific Translation, 1968.

Cathey, H. M., & Campbell, L. E. (1982). Plant response to light quality and quantity. In *Breeding plants for less favorable environments*, ed. M. N. Christiansen & C. F. Lewis, pp. 213–57. New York: Wiley.

de Fossard, R. A. (1976). *Tissue culture for plant propagators*. Armidale (Australia): University of New England.

Gautheret, R. J. (1959). *La culture des tissus végétaux: Techniques et réalisations*. Paris: Masson.

Murashige, T. (1974). Plant propagation through tissue culture. *Annu. Rev. Plant Physiol.* 25, 135–66.

Street, H. E. (1973). Laboratory organization. In *Plant tissue and cell culture*, ed. H. E. Street, pp. 11–30. Oxford: Blackwell Scientific Publications.

Wetherell, D. F. (1982). *Introduction to in vitro propagation*. Wayne, N.J.: Avery Publishing Group.

White, P. R. (1963). *The cultivation of animal and plant cells*, 2d Ed. New York: Ronald Press.

# 3

## Aseptic techniques

The importance of maintaining a sterile environment during the culture of plant tissues is absolutely necessary and cannot be overemphasized. A few simple precautions to avoid contamination will save valuable time in not repeating experiments. The best example of aseptic technique can be found in the operating room of a modern hospital.

The single most important factor in the selection of a suitable working area is the possible flow of unfiltered air over the disinfected working area. Air currents must be avoided because of airborne spores of contaminating microorganisms. An interior room, similar in layout to a photographic darkroom, is an excellent choice for aseptic procedures. Because opening the door creates a draft, post a NO ADMITTANCE sign on the door during aseptic procedures. If precautions are taken, an open laboratory bench can be used in a draft-free room. Under these conditions it may be prudent to use either a face mask or a plastic biohazard shield. White (1963) employed a transparent plastic shield parallel to the top of an open bench to prevent spore fallout.

Most plant scientists using tissue culture procedures conduct sterile operations within some type of transfer chamber, bacteriological glove box, or laminar flow cabinet. With a laminar flow cabinet (Fig. 3.1) air is forced through a dust filter and then passed through a high-efficiency particulate air (HEPA) filter. Depending on the type of cabinet, the air is directed either downward or outward over the working area (Torres, 1989). The gentle flow of sterile air is designed to prevent any spore-laden unfiltered air from entering the cabinet. Since ethanol is highly inflammable, flaming instruments in a laminar flow cabinet should be done with caution: The air flow from the cabinet would direct a flash fire toward the worker (Wetherell, 1982). A few minutes before the start of any sterile procedure, the working area should be thoroughly scrubbed with a tissue soaked with ethanol or isopropanol (70% v/v). Although Dixon (1985) has indicated the use of a 20% aqueous solution of phenol for surface disinfection, volatile phenolics in the laboratory may pose a threat to the growth of tissue cultures (Hamilton, 1973).

Aseptic cabinets and transfer rooms are often equipped with one or more germicidal lamps emitting ultraviolet (UV) light. This type of radiation is useful in eliminating airborne contaminants and for surface disinfection. The emission at 254 nm is slowly germicidal, but UV does not penetrate surfaces; dust and shadowed areas protect contaminants from its effects. Although the effectiveness of UV lamps in creating a sterile environment is questionable (Klein & Klein, 1970; Collins & Lyne, 1984), lamp performance can be tested with the Uvicide Germicidal Lamp Monitor (Vangard International, Inc.). The monitor is the size of a business card and has three pink germicidal-radiation-sensitive dots. One dot is a reference, another is calibrated to kill typical bacteria, and a third is calibrated to destroy typical mold spores. The refer-

Figure 3.1. Laminar air-flow cabinet arranged for culture.

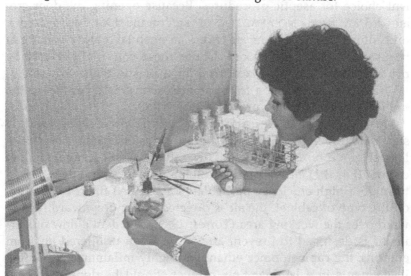

ence dot rapidly changes from pink to red on exposure to 254-nm radiation, whereas the bacteria dot changes color after a dosage of 16,000 $\mu$W-sec/cm$^2$, and the mold spore dot turns red after exposure to 150,000 $\mu$W-sec/cm$^2$. Although this provides some indication of the effectiveness of the lamp, it does not mean complete sterility has been achieved. For example, spores of *Aspergillus niger* and *Rhizopus nigricans* survive dosages of UV radiation considerably higher than 150,000 $\mu$W-sec/cm$^2$ (Westinghouse Electric Corp., 1976). Germicidal lamps have a relatively short life, although they will continue to emit visible radiation after emission at 254 nm has ceased. In the presence of plastics, the radiation may produce inhibitory substances in culture media (Collins & Lyne, 1984). The use of UV radiation should be kept to a minimum since it is not a substitute for cleanliness. If such a lamp is used, switch it on about 30 min prior to culture time. Do not leave the UV lamp on for several hours or overnight because the accumulation of toxic ozone in a confined space creates a hazard.

An extremely important point about aseptic procedure, and one of the leading causes of contamination, is unclean hands. Simply rinsing the hands with water is insufficient; it is necessary to scrub them *vigorously* with soap and hot water for several minutes. Attention must be given to the fingernails and to any part of the forearm that extends into the working area. After a hot-water rinse, blot the skin partially dry with paper towels. It is not advisable to use strong disinfectants that could produce a skin rash. The hands may be dipped in a dilute solution of ethanol or isopropanol, although this can cause skin dryness and must not be used indiscriminately around an open flame.

According to Biondi and Thorpe (1981), the seed coat can be removed manually from surface-sterilized seeds without contamination by dipping the fingers repeatedly in ethanol (40–70% v/v) prior to the operation. Hexachlorophene (2,2´methylenebis [3,4,6-trichlorophenol]) has been used as an antibacterial agent in soaps. This chemical penetrates the intact skin and can cause brain damage. It is not recommended that pHisohex or other preparation containing this chemical be used (Wade, 1971). As an alternative to bare hands, some workers prefer to use disposable sterile surgeon's gloves.

Several techniques are used for the sterilization of glassware, surgical instruments, liquids, and plant material. The term "sterilization" is an absolute one that implies the total inactivation of all forms of microbial life in terms of the ability of the organisms to reproduce (Willett, 1988). "Disinfection," on the other hand, means the reduction of bacterial numbers to some arbitrary "acceptable" level (Hamilton, 1973). Ethanol is a bacteriostatic agent and disinfects a working area, but the treated area is not sterile. The methods of sterilization can be classified as follows: dry heat, wet heat, microwave, microfiltration, and chemical.

*Dry heat.* This method is used only for glassware, metal instruments, and other materials that are not charred by oven temperatures. Strips of tape should not be on any glassware, and the cutting edges of surgical blades may be dulled by the prolonged oven heat. Cotton, paper, and plastic should never be placed in the oven for sterilization. Although laboratory drying ovens may be used, the oven of a gas or electric kitchen stove will serve the same purpose. It is highly important not to use a laboratory oven that has previously been used for paraffin embedding.

Objects to be sterilized are wrapped in heavy-duty aluminum foil before being placed in the oven. Take care not to pack the foil-wrapped packages in the oven too tightly, but to leave some space between them (Hamilton, 1973). In the United States three grades of foil are available. The thinnest grade often has pinholes, and the extra-heavy-duty is too thick. After sterilization the wrapped packages are taken to the laminar flow cabinet. In calculating the time required for dry-heat sterilization, three time periods must be considered:

1. Approximately 1 hr (heating-up period) is allowed for the entire load to reach the sterilization temperature of 180 °C (356 °F).
2. A minimum of 2 hr at this temperature is required to kill all organisms including the spore formers (Willett, 1988).
3. Finally, a cooling-down period is advisable in order to prevent the glassware from cracking due to a rapid drop in temperature.

*Wet heat.* This procedure employs an autoclave operated with steam under pressure (Fig. 3.2). If the laboratory is not equipped with an autoclave, a home pressure cooker can be used. For the sterilization of paper products, glassware, instruments, and liquids the standard procedure is for the autoclave to operate at a steam pressure of 15 lb/in. (103.4 kPa) and a chamber temperature of 121 °C (250 °F). For anything other than liquids, the time required for sterility is 15 min *after* the chamber has reached the sterilization temperature of 121 °C. The time required for the sterilization of liquids varies considerably depending on the volume: The greater the volume, the longer it will take for the contents to reach sterilization temperature. Solid-state temperature probes have given accurate measurements of the time required for various volumes to reach 121 °C (Burger, 1988). The minimum autoclaving times, which includes the time required to reach 121 °C plus 15 min, are as follows ($cm^3$/min): 25/20, 50/25, 100/28, 250/31, 500/35, 1,000/40, and 2,000/48. Unfortunately, there are evidently

Figure 3.2. Portable electric autoclave for wet sterilization of media and equipment. (Courtesy of Gallenkamp.)

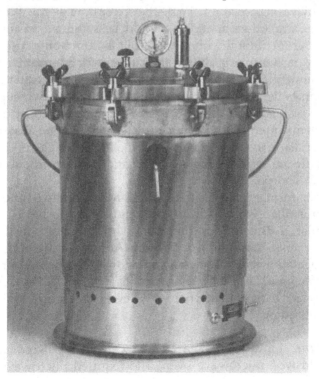

some heat-resistant species of *Bacillus* that can withstand recommended sterilization times as well as instrument flaming (Leifert & Waites, 1990).

All of the residual air in the chamber must be removed so that the steam is in direct contact with all the materials in the autoclave. If a pressure cooker is used, do not close the escape valve until a steady stream of pure steam is evident. At the end of the sterilization period, the pressure must be permitted to return to the atmospheric level slowly because rapid decompression will cause the liquids to boil out of the vessels. Prolonged autoclaving must be avoided because it results in the degradation of certain components of the medium. For example, one group reported that 5% of the sucrose in a liquid medium was hydrolyzed during autoclaving (Hagen et al., 1991). This topic is discussed in Chapter 4. If the autoclave does not have a drying cycle, paper products should be placed in a drying oven (<60°C) briefly in order to evaporate the condensed moisture. Steam in the autoclave chamber must penetrate the materials; a temperature of 121°C will not by itself achieve sterilization. With the exception of the flasks, instruments and other materials should be wrapped in unwaxed kraft paper. Although aluminum foil is commonly used as a wrapping, it is impermeable to the steam vapors and therefore is not recommended (Hamilton, 1973). Demineralized water should be used in boilers of autoclaves that generate their own supply of steam, as well as in pressure cookers. Steam generated by external power plants often contains contaminates that may be absorbed by the materials in the autoclave (Kordan, 1965; Bonga, 1982). Caution should be used in subjecting plastic labware to steam heat (Biondi & Thorpe, 1981).

Another form of wet-heat sterilization is a boiling-water bath. The bath is filled with distilled water and heated to 100°C. Instruments, placed directly into the water, should remain in the boiling water for a minimum of 20 min. Although microorganisms in the vegetative state are destroyed by this treatment, spores will be unaffected (Hamilton, 1973; Collins & Lyne, 1984). Also, a boiling-water bath is ineffective as a device for sterilization above 6,000 ft elevation.

*Microwave.* Liquid and agar media can be sterilized using a household-type microwave oven. The required microwave treatment is somewhat empirical since it depends on the energy produced by the magnetron, vessel type, volume of medium, and the presence of energy-sink water reservoirs. The problem of the agar media boiling over in test tubes was diminished by placing in the oven two 1-liter Pyrex borosilicate glass bottles containing distilled water (Tisserat, Jones, & Galletta, 1992); liquid media can be microwaved without the use of the energy-sink water bottles. A low level of boiling over, coupled with a high rate of sterilization, was achieved with 7.5- and 10-min microwave treatment using the two water bottles each with 250 or 900 ml distilled water. Energy levels of 490 and 700 W gave complete sterilization (Tisserat et al., 1992). These workers reported that the growth of cultured lemon fruits, strawberry organogenic shootlets, and carrot asexual embryogenic callus on microwave-treated media did not differ from growth on autoclaved media. To the best of our knowledge there is no information, to date, on the possible degradation of media components by microwave treatments.

*Microfiltration.* Microfiltration is the process of removing contaminants in the range of 0.025–10 $\mu$m from fluids by passage through a microporous medium, such as a membrane filter. This is necessary for media components that are degraded during autoclaving and must be sterilized at room temperature. A relatively small volume can be sterilized by passage through a filtration unit attached to a hypodermic syringe. For example, the Swinney (stainless steel) and the Swinnex (polypropylene) are reusable units equipped with membrane filters, whereas the Millex units are disposable (Millipore Corporation). Millex units are available in 4-, 13-, 25-, and 50-$\mu$m diameters. The appropriate volume of the sterile liquid is added directly to the autoclaved medium with a graduated syringe. If an agar medium is used, this is done while the agar is still hot (approximately 45 °C) and in the liquid (sol) state. For larger volumes special units are equipped for either vacuum or pressure operation.

Although several different kinds of microfilters are available, their merits and disadvantages are beyond the scope of this introduction. Most of the membrane filters are screens consisting of a uniform continuous mesh of polymeric material with pore size precisely determined by the manufacturing process. It is important to select a filter with a low level of water extractables since the release of these substances can be inhibitory to cell cultures (Cahn, 1967). An excellent choice for plant tissue culture work is the Millipore Durapore (hydrophilic) membrane filter, which contains less than 0.5% by weight of water extractables. For the sterilization of hydrophobic solvents, such as dimethyl sulfoxide solutions, a Fluoropore (Millipore) filter constructed of polytetrafluoroethylene is recommended. Disposable filtration units composed of polystyrene with 0.1- or 0.2-$\mu$m cellulose acetate membranes are practical for infrequent filtrations. These units have been presterilized by radiation and may be purchased with a variety of filter pore sizes and filling and receiving containers. Although some workers use a 0.45-$\mu$m pore size, the pore diameter should measure at least 0.22 $\mu$m for the complete removal of all bacteria, yeasts, and molds (Torres, 1989).

*Chemicals.* The working area is generally disinfected with either ethanol or isopropanol (70% v/v). Although acidified alcohol (70% v/v, pH 2.0) may be a more effective disinfectant, it has a corrosive effect on metal instruments. A higher concentration of ethanol (80% v/v), a more inflammable mixture, is used for flaming instruments. An ethanol dip can be assembled by inserting a large test tube (20 mm O.D. × 150 mm) filled with ethanol (80% v/v) into an empty metal can. After immersion in the alcohol, the instrument is then passed through the flame of a methanol lamp. Avoid prolonged heating of the instrument after the methanol has evaporated. When not in use, the test tube should be capped to retard evaporation of the ethanol.

The surface sterilization of plant material may be accomplished with an aqueous solution of either sodium hypochlorite (NaOCl) or calcium hypochlorite (Ca[OCl]$_2$). Most workers use a common household bleach such as Clorox. (*Do not* use Clorox 2, a recent product whose active ingredient is hydrogen peroxide: It does *not* contain sodium hypochlorite.) These commercial products contain about 5% NaOCl as the active agent. These are usually used at concentrations from 10–20% (v/v), that is, containing about 0.5–1.0% NaOCl. For example, after diluting Clorox with water (1 part bleach:9 parts distilled water), the final concentration will be about 0.5% NaOCl. This concentration is adequate for the surface sterilization of pith parenchyma explants. There is no general agreement in regard to NaOCl concentrations, and higher levels are often used. Tisserat (1985) recommends exposing date palm roots to 2.6% NaOCl for 30 min. Because of complete dissociation, hypochlorite has relatively little activity at pH levels above 8.0, and it is much more effective by buffering at about pH 6.0 (Behagel, 1971). The sterilization may be enhanced by agitating the solution and by the addition of a wetting agent such as Tween 20 or Tween 80.

The following procedure – preparation of a potato tuber for explant removal – may be used as an example. Scrub the unpeeled tuber surface with water containing a detergent. Rinse the tuber under running tap water for a few minutes. Immerse the tuber for 30 min in a 20% bleach solution (200 cm$^3$ bleach:800 cm$^3$ distilled water). Add 10 drops of Tween 20 (polyoxyethylene sorbitan monolaurate) to the bleach solution (Tisserat, 1985). After protecting the hands with surgical gloves, peel the tuber leaving only the inner parenchymatous tissue. Care must be taken to remove *all* traces of eyes (buds), and surface blemishes. Cut rectangular blocks of tissue and surface sterilize them in a 10% bleach solution for 15 min. Gently agitate the solution on an orbital shaker, with a magnetic stirrer, or by hand. The tissue blocks must then be thoroughly rinsed with three changes of sterile DDH$_2$O in order to remove all traces of the hypochlorite. Subsequently, cylindrical primary explants are prepared with a sterile cork borer. A similar procedure was used in studies on potato callus (Hagen et al., 1990). Because of the corrosive effect on metal instruments within the chamber, the bleach solution and the rinse water should be discarded immediately after use.

Several agents were tested by Sweet and Bolton (1979) for the surface sterilization of seeds. Calcium hypochlorite was one of the most effective and least injurious agents. Sodium ions (i.e., in sodium hypochlorite) can induce abnormal development in some seedlings. In addition to Ca(OCl)$_2$, the mixture contained a phosphate buffer giving a final pH of 6.0, and a 1.0% solution of either Triton or Tween 80 as a wetting agent. The seeds were immersed in the hypochlorite mixture for 10 min and then rinsed in three changes of sterile distilled water. Other workers have used 9–10% calcium hypochlorite solutions for periods ranging from 5 to 30 min (*Sigma Plant Cell Culture 1991–92*).

Some workers prefer to treat seed surfaces with a combination of hypochlorite and ethanol. One worker suggests treating barley seeds in the following manner: Seeds were immersed in a 20% bleach solution containing a detergent and agitated 15–20 min with a magnetic stirrer. They were transferred aseptically to a Büchner funnel and rinsed with sterile $DDH_2O$. Then the seeds were covered with 70% ethanol for 1 min, and finally rinsed with sterile $DDH_2O$ (Stiff, 1991).

Some plant tissues pose a problem if they contain microorganisms within the tissue sample. Surface treatments will obviously be completely ineffective. If a fleshy organ examined for explant preparation shows any localized discoloration, it should be rejected. The goal in surface sterilization is to remove all of the microorganisms with a minimum of damage to the plant cells to be cultured. In some cases, the achievement of this goal is empirical and the worker must be flexible in the approach to the problem (de Fossard, 1976). In the case of seeds, the use of higher concentrations of chemicals or longer periods of treatment does not appear to improve the decontamination without reducing the percentage of germination (Sweet & Bolton, 1979).

Leifert and Waites (1990) have reviewed the problem of contaminants of plant tissue cultures. The most difficult plant material for surface sterilization included plant tissues exposed to or near the soil, field-grown plants in tropical climates, and plants receiving overhead irrigation. Faulty sterile technique is a major cause of culture contamination. Approximately 50% of the bacteria found after numerous subcultures are unlikely to have been present in the primary explant. The identification of *Staphylococcus* spp., *Micrococcus* spp., and *Candida albicans* in contaminated cultures clearly signifies human error since these microorganisms are obligate human parasites (Leifert & Waites, 1990).

There are several other agents that have been used as disinfectants, and some are not recommended for student use. These will be discussed later in this chapter, in the section on health hazards.

*Antibiotics.* Although antibiotics are used routinely in animal cell cultures, they have not been widely used in plant tissue cultures (Eichholtz, Hasegawa, & Robitaille, 1982). The early botanists were aware that these natural products may alter the growth and development of plant tissues cultured in vitro (Gautheret, 1959; Butenko, 1964). Falkiner (1990) has reviewed the criteria for the selection of the appropriate antibiotic for the control of bacteria in plant tissue cultures. Antibiotics should be regarded as a form of prophylaxis. Prophylactic use of antibiotics in medicine depends on three conditions, according to Falkiner (1990):

1. A single known pathogen is targeted;
2. the pathogen remains sensitive to the drug; and
3. the period of exposure to the drug is short and limited to the period of maximum risk.

These criteria are helpful, to some extent, to plant scientists. Identification of the contaminants is highly important, and it would be prudent to limit the treatment to the stages of culture when prophylaxis would be most beneficial. The effectiveness of antibiotics is questionable, and their usage is not a substitute for the strict adherence to proper sterile techniques. No known antibiotic is effective against all microorganisms that might cause contamination. In one study eight agents were tested, and none was capable of completely eradicating bacterial contamination (Bastiaens et al., 1983). These agents, or their degradation products, may be metabolized by plant tissues with unpredictable results. Xylogenesis in explants of lettuce pith and Jerusalem artichoke tuber was strongly inhibited in the presence of gentamicin sulfate, although this effect was observed within the concentration range (50–100 $\mu$g/cm) recommended by the manufacturer for use in tissue cultures (Dodds & Roberts, 1981). If antibiotic usage is deemed necessary, the agents of choice are those that act specifically within bacterial cell walls and bacterial membranes. Selection of an effective agent cannot be made until the contaminants are known. Falkiner (1990) has briefly mentioned the merits and possible disadvantages of the recently released drugs.

*Health hazards.* Hypochlorite solutions should be used with care. Inhalation can produce severe bronchial irritation and skin contact can be harmful (Windholz, 1983). One should never use the mouth to pipette a hypochlorite solution, or any other chemical solution used in the tissue culture laboratory: Pipette fillers are available and inexpensive. Never use hypochlorite or other inorganic chloride solution in the presence of UV radiation: The resulting release of chlorine gas is a serious risk (Hamilton, 1973).

A fire danger exists if the student, after flaming an instrument, reinserts the hot instrument into the alcohol dip. Ethanol is highly inflammable. One must be extremely careful about spilling ethanol or other alcohol in the vicinity of an open flame. The test tube containing the 80% v/v ethanol should be kept in a metal can to avoid the release of burning ethanol in case of glass breakage.

Ultraviolet radiation poses some serious health risks. One should never look at a live tube with the naked eye. UV burns to the eye are very painful, although not normally of lasting effect (Scherberger, 1977). A glass barrier between the eyes and the UV source provides complete protection. UV radiation can produce irritation to unprotected skin, so avoid placing the hands in the cabinet when the lamp is on. Another problem is the formation of ozone ($O_3$) resulting from the photochemical reaction with atmospheric oxygen. This explosive gas is a powerful oxidizing agent, and high concentrations can cause severe irritation to the respiratory tract and eyes. Symptoms of ozone toxicity have been experienced by airline passengers and crew during high-altitude flights (Broad, 1979). The UV lamp should never be left on for long periods of time with the hood closed.

Mercuric chloride ($HgCl_2$) has been used for surface sterilization, but this chemical is an *extremely* dangerous poison. An aqueous solution of $HgCl_2$ is slightly volatile at room temperature, and this has resulted in mercury poisoning to laboratory workers.

Although Lysol (3.5% v/v) kills vegetative cells, it is ineffective in eliminating spore contamination (Hamilton, 1973). Lysol is a preparation of cresols and phenol and consequently leaves an oily film on surfaces. In addition, it can produce a severe burn to the skin. If residual Lysol is autoclaved, it will result in the chemical contamination of the contents of the autoclave plus the autoclave itself (Hamilton, 1973). Because Lysol is a common constituent of industrial floor cleaners, one should make certain that such chemicals are not used in the tissue culture laboratory. Volatile phenols can have an adverse effect on cultures (Hamilton, 1973).

Gas sterilization should not be used in a classroom laboratory without *strict* supervision by an instructor trained in the handling of toxic gases. Small plastic objects can be sterilized by subjecting them to a saturated atmosphere of ethylene oxide for several hours in a sealed container at room temperature. Ethylene oxide is violently explosive in nearly all mixtures with air, and the gas is toxic at concentrations not detected by smell (Sykes, 1969). It is highly irritating to the eyes and mucous membranes, and high concentrations can cause pulmonary edema. Basic procedures for its use may be found in the review by Hamilton (1973). In addition, the sterilization of polycarbonate culture flasks with ethylene oxide can result in the formation of some substances that are mutagenic to cultured animal cells (Krell, Jacobson, & Selby, 1979).

Special care must be taken in handling certain antibiotics. Vancomycin and chloramphenicol are toxic to humans, and some of the $\beta$-lactams are allergenic in humans (Falkiner, 1990).

*Avoiding contamination.* One important factor in preventing contamination is the elimination of drafts that carry airborne microorganisms into the working area. Keep all doors and windows closed during aseptic procedures. It is best not to have any other person in the room during aseptic transfers. Avoid breathing or coughing into the working area. A surgical face mask is often worn by those working in laminar flow hoods. It is of the utmost importance that your hands, wrists, and forearms be scrupulously clean, and *never* pass your hand or arm directly over a sterile exposed surface (e.g., a water rinse or an open agar plate). All sterile open surfaces should be placed as far back in the hood as conveniently possible. When pouring sterile liquids, grasp the flask at the base, and keep the hands as far as possible from the open tube or Petri dish receiving the liquid. In opening a sterile Petri dish, hold the lid with the thumb and middle finger on opposite sides, and gently pull the lid back; that is, do not permit the fingertips to pass over the sterile bottom half of the plate. A common practice in microbiology is to flame the mouth of culture tubes and flasks during transfers. Unfortunately, this practice introduces ethylene into the vessels. This gas, a production of combustion, is a plant hormone and therefore can influence the growth and morphogenesis of plant tissue cultures (Beasley & Eaks, 1979). At the conclusion of each step of the procedure, remove all unnecessary glassware, instruments, aluminum foil, and other materials that have been used.

Finally, it is important to remove all contaminated cultures from incubators and plant growth chambers. They will ultimately produce spores and place the entire facility in jeopardy. With a plastic squeeze bottle, add a few cubic centimeters of ethanol (70% v/v) to each of the contaminated cultures.

## QUESTIONS FOR DISCUSSION

1. Can you think of ways of avoiding contamination that were not mentioned in this chapter?
2. What are the most important hazards to your safety as a worker using plant tissue culture techniques?
3. Why is it important to identify the contaminating microorganisms?
4. Can you think of some reasons why antibiotics are not used more commonly in plant tissue cultures?

## APPENDIX

*Closures for tissue culture vessels.* Flasks containing double-distilled water and culture media can be sealed for sterilization with a single layer of heavy-duty aluminum foil. Center the square of foil over the mouth of the flask. Make a circle by pressing the tip of the thumb and index finger together, and with a downward movement press the foil firmly against the neck of the flask. The foil should extend down at least 20–30 cm from the mouth of the flask. The foil cap, however, should *not* be airtight, or it will be blown off during autoclaving by the expansion of the trapped air within the vessel. In addition, it is necessary that the steam come into contact with the medium in order to achieve sterilization.

After placing the explants on the surface of the medium, some workers seal the culture tubes with one or more layers of Parafilm M. This material is placed on top of the previously sterilized aluminum foil cap. Parafilm M is a rubber–wax–polymer mixture approximately 120 $\mu$m in thickness. The manufacturer has informed us that the material will act as an ultrafilter for gas diffusion (i.e., it will impede the flow of gases in and out of culture vessels). The concentration of ethylene produced by certain cultures could possibly rise to inhibitory levels in vessels sealed with Parafilm M.

Polypropylene film has been used as a capping material for tissue culture containers (Mahlberg, Masi, & Paul, 1980). This transparent film, 1 mil or less in thickness, offers certain advantages. Although this film is relatively impermeable to water vapor and prevents the desiccation of long-term cultures, gas exchange is not restricted. The transparency of the film is an advantage for cultures requiring light. Squares of the film cut to an appropriate size are placed individually between the pages of a booklet constructed of ink-free paper. The booklet containing the film is enclosed in an envelope for sterilization by autoclave. Culture tubes, capped in the conventional manner with aluminum foil, are oven sterilized and filled with sterile nutrient medium. The sterile polypropylene is then used to seal the culture tubes after the explants have been positioned on the medium. The original foil caps are discarded. The film is held in place with small rubber bands (Mahlberg et al., 1980).

Several types of commercial culture tube caps are available (Corning; Bellco). Some workers prefer color-coded polypropylene caps, which are autoclavable.

*Surface sterilization of small and hydrophobic plant material.* Some type of carrier system is necessary for minute seeds and small fragments of cutinized plant material that would float on the surface of the hypochlorite solution. Such material can be enclosed in a cheesecloth bag or tea bag. Care must be taken that the bag is securely tied, and the attachment of a string is helpful in moving the specimens through the rinses. A glass rod may be required to hold the bag beneath the surface of the hypochlorite solution. Add a few drops of Tween 20 or other wetting agent to the sterilizing solution. For seeds, a concentration of 15–20% (v/v) NaOCl is recommended, and the seeds should remain submerged for approximately 15 min. For a small bag of specimens 100 cm$^3$ of hypochlorite is adequate, combined with three rinses of 100 cm$^3$ DDH$_2$O each (1 min rinse in each).

SELECTED REFERENCES

Bastiaens, L., Maene, L., Harbaouri, Y., Van Sumere, C., Vande Castelle, K. L., & Debergh, P. C. (1983). The influence of antibacterial products on plant tissue cultures. *Med-Fac. Landbouww. Rijsuniv. Gent. 48/1*, 13–24.

Beasley, C. A., & Eaks, I. D. (1979). Ethylene from alcohol lamps and natural gas burners. Effects on cotton ovules cultured in vitro. *In Vitro 15*, 263–9.

Behagel, H. A. (1971). The pH and sterilization. In *Effects of sterilization on components in nutrient media*, ed. J. van Bragt, D. A. A. Mossel, R. L. M. Pierik, & H. Veldstra, pp. 117–20. Wageningen: H. Veenman & Zonen.

Biondi, S., & Thorpe, T. A. (1981). Requirements for a tissue culture facility. In *Plant tissue culture: Methods and applications in agriculture*, ed. T. A. Thorpe, pp. 1–20. New York: Academic Press.

Bonga, J. M. (1982). Tissue culture techniques. In *Tissue culture in forestry*, ed. J. M. Bonga & D. J. Durzan, pp. 4–35. The Hague: Martinus Nijhoff/Junk.

Broad, W. J. (1979). High anxiety over flights through ozone. *Science 205*, 767–9.

Burger, D. W. (1988). Guidelines for autoclaving liquid media used in plant tissue culture. *HortScience 23*, 1066–8.

Butenko, R. G. (1964). *Plant tissue culture and plant morphogenesis.* Translated from the Russian. Jerusalem: Israel Program for Scientific Translation, 1968.

Cahn, R. D. (1967). Detergents in membrane filters. *Science 155*, 195–6.

Collins, C. H., & Lyne, P. M. (1984). *Microbiological methods.* 5th Ed. London: Butterworths.

de Fossard, R. A. (1976). *Tissue culture for plant propagators.* Armidale (Australia): University of New England.

Dixon, R. A. (1985). Isolation and maintenance of callus and cell suspension cultures. In *Plant cell cultures: A practical approach*, ed. R. A. Dixon, pp. 1–20. Oxford: IRL Press.

Dodds, J. H., & Roberts, L. W. (1981). Some inhibitory effects of gentamicin on plant tissue cultures. *In Vitro 17*, 467–70.

Eichholtz, D. A., Hasegawa, P. M., & Robitaille, H. A. (1982). Effects of gentamicin on growth of shoot initiation from cultured tobacco callus and *Salpiglossis* leaf discs. *In Vitro 18*, 12–14.

Falkiner, F. R. (1990). The criteria for choosing an antibiotic for control of bacteria in plant tissue culture. *Newsl. IAPTC 60*, 13–23.

Gautheret, R. J. (1959). *La culture des tissus végétaux: Techniques et réalisations.* Paris: Masson.

Hagen, S. R., LeTourneau, D., Muneta, P., & Brown, J. (1990). Initiation and culture of potato tuber callus tissue with picloram. *Plant Growth Regulation 9*, 341–5.

Hagen, S. R., Muneta, P., Augustin, J., & LeTourneau, D. (1991). Stability and utilization of picloram, vitamins, and sucrose in a tissue culture medium. *Plant Cell, Tissue & Organ Cult. 25*, 45–8.

Hamilton, R. D. (1973). Sterilization. In *Handbook of phycological methods*, ed. J. R. Stein, pp. 181–93. Cambridge: Cambridge University Press.

Klein, R. M., & Klein, D. T. (1970). *Research methods in plant science.* Garden City, N.Y.: Natural History Press.

Kordan, H. A. (1965). Fluorescent contaminants from plastic and rubber laboratory equipment. *Science 149*, 1382–3.

Krell, K., Jacobson, E. D., & Selby, K. (1979). Mutagenic effect on L5178Y mouse lymphoma cells by growth in ethylene oxide-sterilized polycarbonate flasks. *In Vitro 15*, 326–34.

Leifert, C., & Waites, W. M. (1990). Contaminants in plant tissue cultures. *Newsl. IAPTC 60*, 2–13.

Mahlberg, P. G., Masi, P., & Paul, D. R. (1980). Use of polypropylene film for capping tissue culture containers. *Phytomorphology 30*, 397–9.

Scherberger, R. F. (1977). Ultraviolet radiation, "black" and otherwise. *Eastman Org. Chem. Bull. 49*, 1–2.

*Sigma Plant Cell Culture 1991–92.* St. Louis: Sigma Chemical Company.

Stiff, C. M. (1991). *Barley tissue culture methodology: Media, culture techniques, and approaches to biolistic transformation.* Privately printed by the author.

Sweet, H. C., & Bolton, W. E. (1979). The surface decontamination of seeds to produce axenic seedlings. *Am. J. Bot. 66*, 692–8.

Sykes, G. (1969). Methods and equipment for sterilization of laboratory apparatus and media. In *Methods in microbiology*, ed. J. R. Norris & D. W. Ribbons, vol. 1, pp. 77–121. New York: Academic Press.

Tisserat, B. (1985). Embryogenesis, organogenesis and plant regeneration. In *Plant cell culture: A practical approach*, ed. R. A. Dixon, pp. 79–105. Oxford: IRL Press.

Tisserat, B., Jones, D., & Galletta, P. D. (1992). Microwave sterilization of plant tissue culture media. *HortScience 27*, 358–61.

Torres, K. C. (1989). *Tissue culture technique for horticultural crops.* New York: Van Nostrand Reinhold.

Wade, N. (1971). Hexachlorophene: FDA temporizes on brain-damaging chemical. *Science 174*, 805–7.

Westinghouse Electric Corp. (1976). *Westinghouse Sterilamp Germicidal Ultraviolet Tubes.* Bloomfield, N.J.: Lamp Commercial Division.

Wetherell, D. F. (1982). *Introduction to in vitro propagation.* Wayne, N.J.: Avery Publishing Group.

White, P. R. (1963). *The cultivation of animal and plant cells*, 2d Ed. New York: Ronald Press.

Willett, H. P. (1988). Sterilization and disinfection. In *Zinsser microbiology*, 19th Ed., ed. W. K. Joklik, H. P. Willett, D. B. Amos, & C. M. Wilert, pp. 161–71. Norwalk, Conn.: Appleton & Lange.

Windholz, M. (ed.) (1983). *The Merck index: An encyclopedia of chemicals and drugs*, 10th Ed. Rahway, N.J.: Merck.

# 4

# Media composition and preparation

Because there is a division of labor by different organs of the plant in the biosynthesis of organic metabolites, we know less about the nutritional requirements of the individual organs and tissues of the plant than we do of the whole plant. A tomato plant growing in the garden requires only an external supply of mineral elements for the successful completion of its life cycle. An isolated root of this tomato plant has different requirements for normal growth and development. In addition to the essential mineral elements, the isolated root also requires certain organic compounds because in the whole plant the root system was provided with these compounds. That is, essential organics were synthesized elsewhere in the plant and transported down to the root system. In 1934 White discovered that isolated tomato roots had the potential for unlimited growth if they were provided with a liquid medium containing a mixture of inorganic salts, sucrose, thiamine, pyridoxine, nicotinic acid, and glycine. In addition to the somewhat limited biosynthetic activity of isolated tissues and organs, cultured systems may exhibit changes in their metabolic pathways over a period of time. These changes in metabolism often require corresponding changes in external nutrition. The requirement for a particular organic supplement could be due either to the inability of the culture to produce it or to a new requirement resulting from a shift in metabolism.

The components of a plant tissue culture medium include macronutrients, micronutrients, a separate iron supplement, vitamins, a carbon source, and usually plant growth regulators. Amino acids and various nitrogenous compounds may be present in the vitamin mixture. Other topics discussed in this chapter include complex organic additives, charcoal, osmotica, water, medium matrix, and high-temperature degradation of media components.

*Macronutrients.* Cultured plant tissues require a continuous supply of certain inorganic chemicals. Aside from carbon, hydrogen, and oxygen, the essential elements required in relatively large amounts are termed macronutrient elements. The macronutrients are nitrogen, phosphorus, potassium, calcium, magnesium, and sulfur. Nitrogen, added in the largest amount, is present as either a nitrate ($NO_3^-$) or an ammonium ($NH_4^+$) ion, or a combination of these ions. Magnesium sulfate ($MgSO_4 \cdot 7H_2O$) satisfies both the Mg and S requirements. Sulfur may also be present in the form of $Na_2SO_4$. Phosphorus can be represented by $NaH_2PO_4 \cdot H_2O$, $KH_2PO_4$, or $(NH_4)H_2PO_4$. Potassium, the cation found in the largest amount, is given as KCl, $KNO_3$, or $KH_2PO_4$. The calcium requirement involves $CaCl_2 \cdot 2H_2O$, $Ca(NO_3)_2 \cdot 4H_2O$, or anhydrous forms of these salts.

*Micronutrients.* Traces of certain mineral elements are required by all plant cells. Because these quantities are exceedingly small for some of the elements, a more concentrated stock solution (except for iron) is prepared in advance. (It is also convenient to prepare in advance stock solutions of macronutrients, vitamins, and certain plant growth regulators. These solutions can be stored for limited periods of time in glass containers at 4 °C.) Micronutrient elements essential for all higher plant cells include iron, manganese, zinc, boron, copper, molybdenum, and chlorine (Marschner, 1986).

Certain plants have unique requirements for some micronutrients; media nurturing explants taken from these plants could have similar requirements. Sodium could be a micronutrient for the culture of certain halophytes, plants with $C_4$ photosynthetic pathways, and plants with Crassulacean acid metabolism. Since nickel is essential for the structure and functioning of urease (Klucas, Hanus, & Russell, 1983), cultures of jackbean (*Canavalia ensiformis* L.) and soybean (*Glycine max* L.) may require a trace of nickel in the medium. Although iodine in the form of KI is a constituent of several media, the necessity of this element remains questionable. A trace of cobalt is found in several media, and yet this element is not known to have any function in higher plants. One unresolved problem is that even the purest chemical reagents will contain traces of inorganic contaminants, and these always constitute a hidden source of micronutrients (Yeoman, 1973). Agar is a source of numerous minerals, possibly traces of vitamins, and even toxic substances (Pierik, 1971). Little critical work has been done on the micronutrient requirements of media designed for special purposes.

There are inaccuracies and errors in the molecular formulae of the inorganic components of several media and inconsistencies in citations subsequent to the original papers. Owen and Miller (1992) have published the correct concentrations and chemical formulations of inorganic constituents for the following media: White (1943, 1963), Murashige and Skoog (1962), B5 (Gamborg, Miller, & Ojima, 1968), Nitsch and Nitsch (1969), Chu (1978), Chu et al. (1975), Anderson (1980), and Lloyd and McCown's (1980) woody plant medium. Since some commercial basal media preparations have inconsistencies, the formulations provided by the companies should be checked carefully (Owen & Miller, 1992).

*Iron supplement.* An iron stock is prepared separately because of the problem of iron solubility. This element requires acidic conditions for solubility. Usually the iron stock is prepared in a chelated form as the sodium salt of ferric ethylenediaminetetraacetic acid (NaFeEDTA). EDTA itself, however, may have side effects on certain enzyme systems and on morphogenesis in cultures (Bonga, 1982). Some inhibitory effects may be due to light-induced degradation of EDTA (Hangarter & Stasinopoulos, 1991). Further information is available on the preparation of ferric salts of EDTA (Steiner & van Winden, 1970). Additional work is needed in finding a suitable nonreactive carrier for iron for use in culture media.

*Vitamins.* Vitamins have catalytic functions in enzyme systems and are required only in trace amounts. Thiamine (vitamin $B_1$) may be the only essential vitamin for nearly all plant tissue cultures, whereas nicotinic acid (niacin) and pyridoxine (vitamin $B_6$) may stimulate growth (Gamborg et al., 1976; Ohira, Ikeda, & Ojima, 1976). Thiamine is added as thiamine-HCl in amounts varying from about 0.1 to 10.0 mg/l. The need for thiamine is particularly evident at low levels of cytokinins. In the presence of fairly high concentrations of cytokinins (0.1–10.0 mg/l), tobacco cells grew without the addition of exogenous thiamine (Digby & Skoog, 1966; Linsmaier-Bednar & Skoog, 1967). Presumably tobacco cultures can develop the capability of synthesizing thiamine (Dravnicks, Skoog, & Burris, 1969).

Both nicotinic acid and pyriodixine are required for the culture of *Haplopappus gracilis* (Eriksson, 1965). Nicotinic acid is an essential growth factor for the optimal growth of sugarcane suspension cells (Veith & Komor, 1991).

Some other vitamins that have been used in plant tissue culture media include *p*-aminobenzoic acid (PABA; vitamin $B_x$), ascorbic acid (vitamin C), tocopherol (vitamin E), biotin (vitamin H), choline chloride, cyanocobalamin (vitamin $B_{12}$), folic acid (vitamin $B_c$), calcium pantothenate, and riboflavin (vitamin $B_2$) (Huang & Murashige, 1977; Gamborg & Shyluk, 1981).

*Carbon sources.* All plant tissue culture media require the presence of a carbon and energy source. Sucrose or D-glucose is usually added in concentrations of 20,000–30,000 mg/l. Nearly all cultures appear to give the optimum growth response in the presence of the disaccharide sucrose, whereas there can be considerable variability in growth when other disaccharides or monosaccharides are substituted for sucrose. There are, however, some photoautotrophic cultures that are grown in the absence of an exogenous organic carbon source. In these cultures $CO_2$ fixation via photosynthesis provides sufficient carbohydrate. Although many laboratories autoclave sucrose with the remainder of the nutrient medium, sucrose is heat labile, and the result is a combination of sucrose, D-glucose, and D-fructose. Such an autoclaved medium may give completely different results compared to a medium containing filter-sterilized sucrose (Ball, 1953).

The cyclitol *myo*-inositol is added to some vitamin supplements as a growth factor at a concentration of 100 mg/l, although higher concentrations may be present in special media. Although this compound is a carbohydrate, it has special functions mainly in the form of phosphoinositides and phosphatidylinositol. Inositol bisphospholipids may have a role in the calcium messenger system (Poovaiah, Reddy, & McFadden, 1987) and the IAA–*myo*-inositol conjugate is thought to play roles in the storage, transport, and release of auxin (Bandurski, 1984). Information on inositol metabolism in plants can be found in the monograph by Morré, Boss, & Loewus (1990).

The choice and concentration of the sugar to be used depend mainly on the tissue to be cultured and the purpose of the experiment. A recent study demonstrated that the optimum concentration of sucrose for the induction of xylogenesis in lettuce pith explants was 0.2% (w/v), although xylem formation was stimulated by the addition of as little as 0.001% (1.0 mg/100 cm$^3$) sucrose (Warren Wilson et al., 1994).

The question of the purity of the carbohydrate reagents has been raised because the occlusion of a variety of organic substances, particularly traces of amino acids, occurs during the crystallization of sucrose (Schneider, Emmerich, & Akyar, 1975). Street (1969) summarized some of the tissue culture studies utilizing various carbohydrate sources.

*Plant growth regulators.* The growth regulator requirements for most callus cultures are some combination of auxin and cytokinin (Fig. 4.1). The term "hormone" should be reserved for naturally occurring plant growth regulators. Synthetic growth regulators, such as 2,4-D and kinetin, are not considered to be plant hormones. Auxins, a class of compounds that stimulate shoot cell elongation, resemble IAA in their spectrum of activity. As a supplement they are useful to stimulate the formation of adventitious roots, inhibit bud formation, and to play a role in embryogenesis. Cytokinins, which promote cell division in plant tissues under certain bioassay conditions and only in the presence of auxin, regulate growth and development in the same manner as kinetin (6-furfurylaminopurine). In addition to callus initiation, cytokinins stimulate bud proliferation and inhibit rooting. Cytokinins are mainly $N^6$-substituted aminopurine derivatives, although there are some exceptions.

Auxin–cytokinin supplements are instrumental in the regulation of cell division, cell elongation, cell differentiation, and organ formation. Gibberellins are rarely added to culture media, although $GA_3$ has been used in apical meristem cultures (Morel & Muller, 1964) and studies on vascular differentiation (Roberts, 1988). Increasing attention has been given to ethylene in the initiation of buds (Thorpe, 1982) and tracheary element differentiation

(Miller & Roberts, 1984). Relatively few studies have used abscisic acid as a supplement. Some auxins employed in culture media include IAA, α-NAA, 2,4-D, and picloram. Both the α and β isomers of NAA are commercially available, but the α isomer is always used in media. The β isomer is a weak auxin with relatively little physiological activity. Indole-3-butyric acid (IBA), p-chlorophenoxyacetic acid (4-CPA), and 2,4,5-trichlorophenoxyacetic acid (2,4,5-T) are also effective auxins. IBA is a particularly effective rooting agent. IAA is a naturally occurring auxin, but unfortunately it is rapidly degraded by light (Nissen & Sutter, 1990; Stasinopolis & Hangarter, 1990) and enzymatic oxidation. Because IAA oxidase may be present in cultured tissues, IAA is added to media in a relatively high concentration (1–30 mg/l). The synthetic α-NAA is not subject to the same enzymatic oxidation as IAA, and it may be

Figure 4.1. Structural formulae of some auxins and cytokinins. Auxins include (a) indole-3yl-acetic acid (IBA), (b) α–naphthaleneacetic acid (α-NAA), and (c) 2,4-dichlorophenoxyacetic acid (2,4-D). Cytokinin activity is shown by (d) adenine, (e) kinetin, and (f) *trans*-zeatin.

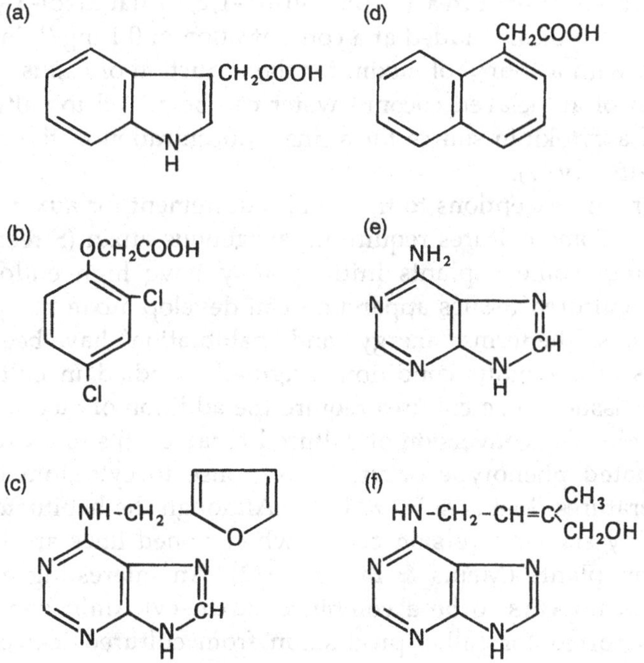

effective in a lower concentration (0.1–2.0 mg/l). One of the most effective auxins for callus proliferation is 2,4-D ($10^{-7}$–$10^{-5}$ M), often employed in the absence of any exogenous cytokinin. This herbicide is a powerful suppressant of organogenesis and should not be used in media involving root and shoot initiation (Gamborg et al., 1976). Picloram (4-amino-3,5,6-trichloropicolinic acid) offers certain advantages over 2,4-D: It is water soluble, effective at lower concentrations than 2,4-D, may be less toxic to plant tissue cultures at optimum levels, and offers the potential for direct regeneration of plants from calli (Collins, Vian, & Phillips, 1978).

The most widely used cytokinins in media are kinetin, 6-benzylaminopurine ($N^6$-benzyladenine), and zeatin (6-[4-hydroxy-3-methyl-but-2-enylamino]purine). Kinetin and 6-benzylaminopurine are synthetic compounds, whereas zeatin occurs naturally in plants. Other naturally occurring cytokinins, which are considerably less expensive than zeatin, are 6-[$\gamma,\gamma$-dimethylallylamino]purine (2iP) or $N^6$-[$\Delta^2$-isopentyl]adenine (IPA). Also, 1,3-diphenylurea exhibits cytokinin-like responses in some bioassays. Another chemical with potent cytokinin activity is thidiazuron (TDZ), a substituted phenylurea ($N$-phenyl-$N'$-1,2,3-thiadiazol-5-ylurea). Kinetin is typically added at a concentration of 0.1 mg/l, in combination with a source of auxin, for the induction of callus. A preparation of autoclaved coconut water can be added to culture media as a cytokinin source for a final concentration in the medium of 10–15% (v/v).

There are exceptions to this dual requirement for auxin and cytokinin. Some cultures require no exogenous auxin (Street, 1966). Although some explants initially may have high endogenous auxin, cultured tissues apparently can develop auxin biosynthetic pathways. The terms "anergy" and "habituation" have been given to this autonomous condition, originally studied in cultures of tumor tissue. Some cultures require the addition of auxin, but not cytokinin. The conversion of cultured tobacco cells to a cytokinin-habituated phenotype occurs in response to cytokinin or high temperatures (Meins & Lutz, 1980). Although the habituated state is highly stable, reversion occurs when cloned lines are induced to form plants (Meins & Binns, 1982). An interesting example of what appears to be a combined auxin–cytokinin habituation was reported for callus production from cultured leaf disks of sugar beet (*Beta vulgaris* L.) cultivars. Callus was initiated on a hormone-free medium after an average of 96.7 days, and subsequently organogenesis occurred on the same medium in some populations (Doley & Saunders, 1989).

*Amino acids and other nitrogenous additives.* With the exception of glycine (aminoacetic acid), which is a component of several media, amino acids are not generally added to plant nutrient media. If a mixture of organic nitrogen is necessary, the medium can be enriched with either casein hydrolysate or casamino acids (0.05–0.1% w/v). Casein, a bovine milk protein, consists of an ill-defined mixture of a least 18 different amino acids (Klein & Klein, 1970). Assuming this supplement has a beneficial effect, additional experiments can be made substituting various amino acids and amides for the hydrolysate. Ultimately one may be successful in identifying the specific organic nitrogen requirement. The composition of different mixtures of amino acids and amides that have been used in plant media are given by Huang and Murashige (1977). Some of the nitrogen compounds that are used most frequently include L-aspartic acid, L-asparagine, L-glutamic acid, L-glutamine, L-arginine, and L-tyrosine. Traces of L-methionine, added to the medium for the enhancement of ethylene biosynthesis, have a stimulatory effect on xylogenesis (Roberts & Baba, 1978; Miller & Roberts, 1984). Often growth inhibition occurs following the addition of a combination of amino acids, a phenomenon that has been attributed to competitive interactions among the various amino acids (Street, 1969).

*Complex organic supplements.* The trend in plant tissue culture has been to attempt to define all of the constituents of a given medium and to eliminate the use of crude natural extracts. Such products as peptone, yeast extract, and malt extract are seldom used today. Although this attitude is commendable from a scientific viewpoint, the use of natural extracts should not be ignored when chemically defined media fail to produce the desired results. For this reason, a simple procedure for the preparation of coconut water from fresh coconuts will be found in the appendix. Fruit juices are also useful supplements. Explants from several *Citrus* spp. were stimulated in growth by the addition of orange juice to the medium (Einset, 1978). Tomato juice (30% v/v) has been used effectively (Nitsch & Nitsch, 1955; Straus, 1960). Banana powder and coconut water are available commercially from Sigma.

*Charcoal.* Activated charcoal (AC) will adsorb many organic and inorganic molecules from a medium (Mattson & Mark, 1971). Although the precise effects of AC are unknown, there are several possible modes of operation. It may remove contaminants from agar (Kohlenbach & Wernicke, 1978) and secondary products secreted by the cultured tissues (Wang & Huang, 1976; Fridborg et al., 1978), or regulate the supply of growth regulators (Weatherhead, Burdon, & Henshaw, 1978). Some of the effects of AC may be due to darkening of the support matrix and thus approximate soil conditions more closely (Proskauer & Berman, 1970). As a nutrient supplement AC has been reported to stimulate embryogenesis (Kohlenbach & Wernicke, 1978). On the other hand, AC can inhibit growth and morphogenesis (Constantin, Henke, & Mansur, 1977; Fridborg et al., 1978). The type of AC used is important because the adsorptive characteristics and pH are dependent on the manufacturing process (Bonga, 1982). Wood charcoal is much higher in carbon content in comparison to bone charcoal, and the latter preparation contains ingredients that may adversely affect plant tissue cultures (Bonga, 1982). Further information on the effects of charcoal-supplemented media can be found in Chapters 8 and 11.

*Osmotica.* The uptake of water by plant cells is governed by the relative water potential values between the vacuolar sap and the external medium. The major components of the nutrient medium that influence water availability are the concentration of the agar, the amount of sugar present, and any nonmetabolite added as an osmoticum. One colloidal characteristic of the gel state of agar is the imbibitional retention of water within the micelles of the gel (Levitt, 1974). Carbohydrates not only function as a carbon source, but they play an important role in the regulation of the external osmotic potential. Often a weakly metabolized sugar – for example, mannitol or sorbitol – is used as an external osmoticum (Brown, Leung, & Thorpe, 1979; Brown & Thorpe, 1980). Mannitol, however, is metabolized by *Fraxinus* cultures (Wolter & Skoog, 1966) and possibly others. Polyethylene glycol (PEG) has been used as an osmoticum in protoplast fusion experiments and in the cryopreservation of cultures. Some additional information on the use of osmotica can be found in the review by Bonga (1982).

*Water.* The water used in tissue culture media, including the water employed in the preparation of the explant, should be double distilled or of equivalent purity. In addition to glass distillation, other water purification methods include screen and depth filters, electrodialysis, carbon adsorption, resin-based deionization, ultrafiltration, and reverse osmosis. Only distillation and reverse osmosis have the capability of removing all the sources of contamination (inorganics, organics, bacteria, pyrogens, and particles). Reverse-osmosis water is of sufficient purity that it is widely used for hemodialysis (IonPure, subsidiary of Millipore). Water distillation is a complex process, and volatile organics of low molecular weight often are distilled along with the water. Bonga (1982) suggests that the distilled water collected during the first 10–15 min of still operation should be discarded in order to eliminate the early vaporization of some of these volatile organics. The use of an ion exchange column poses some technical problems in water purification because of the release of a variety of organic contaminants, including some metabolic products secreted by microorganisms growing within the column (Bonga, 1982). One should be cautioned against the prolonged storage of redistilled water in polyethylene containers since these receptacles release substances that may be toxic to the cultures (Robbins & Hervey, 1974). It is unwise to store double-distilled water in Pyrex vessels for prolonged periods because detectable amounts of bacteria may accumulate during storage under nonsterile conditions (Street, 1973). The lengthy storage of sterile water is unwise, and this has been a problem in some of our hospitals (Favero et al., 1971). Tissue culture water (sterile filtered) is commercially available in 100-cm$^3$, 500-cm$^3$, and 1-gal volumes (Sigma).

*Medium matrix.* Unless the culture is suspended in an aqueous medium, it is grown on a semisolid gel or solid (porous) matrix. Many early experiments were based on agars from Difco Laboratories, and these evidently contained unknown contaminants (Pierik, 1971); consequently, the matrix itself was a nutritional supplement. Today we have a wide range of relatively pure gelling agents. The choice of the appropriate gelling agent, and the concentration to be employed, may be as important as the selection of the ideal nutrient mixture. Here are a few specific examples: An agarose preparation termed Sea Plaque (FMC Corporation) is a suitable matrix. Phytagel (Gelrite, Merck) is an agar substitute synthesized from gellan gum. Agargel (Sigma), a blend of agar and Phytagel, was devised to control vitrification of cultures. Transfergel (Sigma) is a carrier gel, supplemented with a complete nutritional medium, for nurturing propagules such as somatic embryos, microcuttings, and shoot tips. In addition, Sigma offers seven different agar gelling agents suitable for plant tissue culture. It appears that future projects should involve preliminary tests on several gelling agents before making a final choice on the one that produces the best results.

In addition to gels, other materials have been tried. Filter-paper platforms were introduced by Heller (1965), and filter-paper disks impregnated with nutritives have been used (Phillips & Dodds, 1977). Glass fiber filters are useful supports for cultures, although they should be pretreated to remove contaminants and to saturate the cation exchange sites on the glass fibers with specific ions (Tabor, 1981). A synthetic polyester fleece (Pellon Corporation) was used in the culture of Douglas fir plantlets (Cheng & Voqui, 1977). Filter paper has been used to separate contiguous cultures of two different origins, that is, as a "nurse" culture. To measure the growth of a culture with a minimum of disturbance, a thin layer of cells can be separated from the agar medium by a filter-paper disk. The disk and the cultured cells can be periodically removed, weighed aseptically, and replaced on the medium without sacrificing the cells (Horsch, King, & Jones, 1980).

*High-temperature degradation of media components.* Several chemicals employed in plant tissue culture media degrade on exposure to steam sterilization. Gibberellins are rapidly degraded by high temperatures, and the biological activity of a freshly prepared solution of $GA_3$ was reduced by more than 90% as a result of autoclaving (Bragt & Pierik, 1971). The auxins IAA, $\alpha$-NAA, and 2,4-D are relatively thermostable depending on the inorganic basal medium (Dunlap, Kresovich, & McGee, 1986) and supplements (Nissen & Sutter, 1990). Aqueous solutions of kinetin, zeatin, and 2iP have been chromatographed on thin-layer silica-gel chromatograms before and after prolonged autoclaving with no breakdown products detected (Dekhuijzen, 1971). On the other hand, biologically inactive 1,3- or 9-substituted purine molecules were converted into callus-inducing $N^6$-substituted purines by autoclave treatment. Crude plant extracts that possibly contain inactive purine molecules should be filter sterilized. Heat sterilization apparently has no effect on the isomers of abscisic acid (Wilmar & Doornbos, 1971).

Vitamins have varying degrees of thermolability. Most workers autoclave the vitamins with the remainder of the medium. Nicotinic acid, pyridoxine, and thiamine in an MS liquid medium (pH 5.5–5.6) showed no signs of degradation (HPLC analyses) after autoclaving (Hagen et al., 1991); nevertheless, thiamine is rapidly destroyed if the pH of the medium is much above 5.5 (Windholz, 1983). Calcium pantothenate cannot be autoclaved without destruction. If the research study involves vitamin activity, then the vitamins should be sterilized by microfiltration (Ten Ham, 1971). Since those employed in most of our experiments are apparently thermostable, the vitamin supplement will be added to the medium prior to autoclaving.

One of the most frequently employed carbohydrates in media is sucrose. This disaccharide decomposes, to some extent, on autoclaving to release a mixture of D-glucose and D-fructose (Ball, 1953). A recent study indicated that 5 percent of the sucrose in an MS liquid medium was hydrolyzed during autoclaving (Hagen et al., 1991). This degradation can be inhibitory to some cultured tissues (Stehsel & Caplin, 1969; Wright & Northcote, 1972). Presumably the toxicity is due to the degradation products of D-fructose (Rédei, 1974). Steam sterilization may also catalyze reactions within the media between carbohydrates and amino acids (Peer, 1971).

*Additional comments on degradation.* Illumination furnished by cool-white fluorescent tubes promoted the degradation of IAA and IBA in both liquid and agar media (Nissen & Sutter, 1990). Fluorescent lighting in the 290–450-nm band was responsible for the degradation of media components, and this was prevented by the use of yellow long-pass filters (Stasinopoulos & Hangarter, 1990). Growth inhibition of *Arabidopsis* roots in a hormone-free medium was traced to the fluorescent light degradation of EDTA. A photochemical reduction of $Fe^{+3}$ resulted in EDTA oxidation, the release of inhibitory levels of formaldehyde, and the precipitation of iron. The photooxidation of EDTA was prevented by using a yellow acrylic filter to remove the UV and blue wavelengths (Hangarter & Stasinopoulos, 1991).

*Some suggestions on the selection of medium.* The choice of a particular medium depends on the species of the plant, the tissue or organ to be cultured, and the purpose of the experiment. If the plant material has been cultured successfully in other laboratories, it is always best to start with published information. A suitable starting point for the initiation of callus from a dicot tissue explant would be the preparation of the MS basal medium. One characteristic of this medium is its relatively high concentration of nitrate, potassium, and ammonium ions in comparison with other formulations. The B5 medium (Gamborg et al., 1968) is another effective basal mixture. In addition to the basal mineral salts, it is recommended to add the MS vitamin mixture, *myo*-inositol (100 mg/l), and sucrose (2–3% w/v). A possible modification of the MS vitamin mixture would be to increase the thiamine content. The B5 medium contains 10 mg/l thiamine in comparison to 0.1 mg/l in the MS medium. For callus formation the addition of 2,4-D (0.2–2.0 mg/l) serves as an effective auxin, and the addition of kinetin or 6-benzylaminopurine (0.5–2.0 mg/l) is advised. Dixon (1985) has given a variety of plant species that have produced callus on a Schenk and Hildebrandt (1972) medium supplemented with 4-CPA ($10^{-5}M$), 2,4-D ($2 \times 10^{-6}M$), and kinetin ($5 \times 10^{-7}M$). If these combinations fail to produce the desired result, then a supplement of amino acids or some natural plant extract might be considered.

Aside from considerations of callus formation, the initiation of various morphogenetic events in vitro often requires special adjustments in the concentration of the components of the medium. Embryogenesis responds favorably to high levels of potassium in some systems (Brown, Wetherell, & Dougall, 1976). In addition to the proper ratio of growth regulators, shoot formation may require a medium either high in phosphate (Miller & Murashige, 1976) or low in ammonium nitrate (Pierik, 1976). The total salt concentration may be a factor of some importance (Bonga, 1982).

## QUESTIONS FOR DISCUSSION

1. List the inorganic chemicals that you would select for the macronutrient requirements of a typical dicot callus culture. Give the

approximate concentrations that would be suitable for a medium maintaining callus growth.
2. What are some sources of inorganic and organic contaminants that are unwittingly added to our cultures?
3. What evidence indicates that cultured plant tissues can experience changes in certain biochemical pathways over time?
4. What are some possible substitutes for agar as a medium matrix? Discuss their possible advantages and disadvantages.

## APPENDIX

### Preparation of coconut water

Coconuts can be purchased from local fruit markets, and in most cases have been dehusked. Market coconuts will contain about 100–200 cm$^3$ of liquid endosperm. Three micropyles ("eyes") are located at one end of the coconut, one of which is composed of relatively soft tissue, easily removed with a cork borer. The liquid, however, will pour more rapidly if an air vent is created by penetrating a second micropyle with an electric drill or hammer and nail. Each coconut should be drained separately because occasionally one may open a coconut whose liquid has fermented. (These are easily identified because of odor and appearance.) The collected liquid is filtered through several layers of cheesecloth. Boil the filtrate for approximately 10 min in order to precipitate the proteins. Cool to room temperature, decant, and filter the supernatant through a fairly rapid qualitative filter paper. For the preparation of 1 liter of nutrient medium, add 100–150 cm$^3$ of coconut water. Any unused coconut water can be frozen for use at a later date: Melting and refreezing apparently does not diminish the cytokinin-like properties of the substances in the liquid endosperm (Klein & Klein, 1970). For additional information, see Riopel (1973).

### How to prepare stock solutions

Prior to the preparation of a stock solution, always ask yourself the following question: How much of the chemical must be added to the quantity of the medium required? Let us say that 250 cm$^3$ of a nutrient medium requires the addition of 0.1 mg/l kinetin. The latter concentration is equivalent to 0.01 mg/100 cm$^3$ or 0.025 mg/250 cm$^3$. Let us deliver this amount in a 1.0-cm$^3$ pipette; that is, the pipette will contain a kinetin solution of 0.025 mg/cm$^3$. If we simply scale this up, it is the same as 0.25 mg/10 cm$^3$, 2.5 mg/ 100 cm$^3$, and 25 mg/1,000 cm$^3$. Choose a convenient amount of kinetin to weigh out (e.g., 2.5 mg), dissolve the chemical in a few drops of 1 $N$ HCl, and bring the final volume to 100 cm$^3$ with DDH$_2$O in a volumetric flask.

## Solvents for plant growth regulators

Often the choice of a proper solvent for a given plant growth regulator is perplexing. Cytokinins are readily soluble in 1 $N$ HCl. The indole auxins and $\alpha$-NAA can be dissolved in 1 $N$ NaOH. Although 2,4-D is soluble in ethanol, the use of dimethyl sulfoxide (DMSO) is recommended. This powerful solvent has been used previously in several physiological studies (Schmitz & Skoog, 1970; Delmer, 1979; Roblin & Fleurat-Lessard, 1983). At high concentrations, however, DMSO may have adverse effects on metabolism (Vannini & Poli, 1983). Caution must be exercised in the handling of DMSO since it readily penetrates the skin and may have toxic effects (Windholz, 1983).

## Preparation of Murashige and Skoog (MS) stocks

The formulation of Murashige and Skoog's (1962) medium is given in Table 4.1. Additional formulations are given at the end of the book (see "Formulations of tissue culture media"). The MS basal salt mixture is commercially available in powder form from the list of suppliers at the end of the book.

*Iron stock (20x; Table 4.1B).* Dissolve FeSo$_4$·7H$_2$O in 40 cm$^3$ of warm DDH$_2$O in a 100-cm$^3$ beaker. In a separate beaker dissolve Na$_2$EDTA·2H$_2$O in 40 cm$^3$ of warm DDH$_2$O. Mix the two solutions and transfer to a 100-cm$^3$ volumetric flask. Add DDH$_2$O to the final volume. The iron stock should be protected from light by storing the solution in an amber bottle, or wrap the entire flask with aluminum foil (Smith, 1992). Store at room temperature since precipitation may occur at chilling temperatures. Pipette 5 cm$^3$ of iron stock for 1 liter of MS medium.

*Micronutrient stock (100x; Table 4.1C).* Add approximately 400 cm$^3$ DDH$_2$O to a 1-liter beaker. Weigh and dissolve each of the salts

Table 4.1. *Medium for* Nicotiana tabacum *stem callus*

| | Ingredient | Concentration[a] Stock | MS medium |
|---|---|---|---|
| (A) | Macronutrients | | (mg/l) |
| | $(NH_4)NO_3$ | | 1,650 |
| | $KNO_3$ | | 1,900 |
| | $CaCl_2 \cdot 2H_2O$ | | 440 |
| | $MgSO_4 \cdot 7H_2O$ | | 370 |
| | $KH_2PO_4$ | | 170 |
| (B) | Iron | (mg/100 cm$^2$) (20×) | (mg/l) (5 cm$^3$ stock gives) |
| | $Na_2EDTA \cdot 2H_2O$ | 744 | 37.2 |
| | $FeSO_4 \cdot 7H_2O$ | 556 | 27.8 |
| (C) | Micronutrients | (mg/l) (100×) | (mg/l) (10 cm$^3$ stock gives) |
| | $MnSO_4 \cdot 4H_2O$ | 2,230 | 22.3 |
| | $ZnSO_4 \cdot 7H_2O$ | 860 | 8.6 |
| | $H_3BO_3$ | 620 | 6.2 |
| | KI | 83 | 0.83 |
| | $Na_2MoO_4 \cdot 2H_2O$ | 25 | 0.25 |
| | $CuSO_2 \cdot 5H_2O$ | 2.5 | 0.025 |
| | $CoCl_2 \cdot 6H_2O$ | 2.5 | 0.025 |
| (D) | Vitamins | (mg/100 cm$^3$) (100×) | (mg/l) (1 cm$^3$ stock gives) |
| | glycine | 200 | 2.0 |
| | nicotinic acid | 50 | 0.5 |
| | pyridoxine·HCl | 50 | 0.5 |
| | thiamine·HCl | 10 | 0.1 |
| (E) | Auxin | | (mg/l) |
| | picloram | | 3.0 |
| | | | (mg/l) |
| | *myo*-inositol | | 100 |
| | sucrose | | 30,000 |
| | Gelrite (0.2% w/v) | | |
| | pH 5.7 | | |

[a] In the source publication, ranges of concentrations were used for IAA (1–30 mg/l) and kinetin (0.04–10mg/l). Since picloram (4-amino-3,5,6-trichloropicolinic acid) will be the sole plant growth regulator for the first experiment involving potato callus (see Chapter 5), it will be used in this formulation instead of IAA and kinetin. A casein hydrolysate preparation was given as optional (1.0 mg/l). Corrections have been made involving changes in water of hydration for $Na_2EDTA$ and $ZnSO_4$ as given by Owen & Miller (1992).
*Source:* Murashige & Skoog (1962).

given in the first column using a magnetic stirrer. Transfer the solution to a 1-liter volumetric flask, and add $DDH_2O$ to the final volume. Store under refrigeration. Pipette 10 $cm^3$ of the micronutrient stock for 1 liter of MS nutrient medium.

*Vitamin stock (100×; Table 4.1D).* Add about 50 $cm^3$ $DDH_2O$ to a 100-$cm^3$ beaker. Weigh and dissolve each of the vitamins indicated. Transfer the vitamin mixture to a 100-$cm^3$ volumetric flask, and add $DDH_2O$ to the final volume. Store under refrigeration. Pipette 1 $cm^3$ of vitamin stock for 1 liter of MS medium.

*Note:* Do not pipette directly from stock bottles, and do not return any unused stock solutions to the stock bottles. Label all stock solutions and include the concentration, your initials, and the date of preparation. Although inorganic salts are relatively stable in solution under refrigeration, vitamin stock should be discarded after 30 days. Also, vitamin stock should be visually examined periodically for any signs of microorganisms.

## Preparation of the complete MS medium

1. Add approximately 400 $cm^3$ $DDH_2O$ to a 1-liter beaker. Weigh and dissolve each of the macronutrient salts given in Table 4.1A using a magnetic stirrer.
2. Pipette the following from the stock solutions: 5 $cm^3$ iron, 10 $cm^3$ micronutrients, and 1 $cm^3$ vitamins.
3. Weigh 100 mg *myo*-inositol and dissolve it in the medium mixture.
4. Weigh 3.0 mg picloram, dissolve in a few drops of $DDH_2O$, and transfer it to the medium mixture.
5. Add $DDH_2O$ until the total volume of liquid is about 800 $cm^3$. While agitating the solution with a magnetic stirrer, adjust the pH to 5.7 with droplets of 1 N NaOH or 1 N HCl with separate Pasteur pipettes.
6. Transfer the medium to a 1-liter volumetric flask and add $DDH_2O$ to the final volume. Store under refrigeration. Label, initial, and give the date of preparation.

## Final procedure

7. Sterilize the Petri dishes or culture tubes in advance with dry heat (Chapter 3). Several different sizes of culture tubes are avail-

able. Glass shell vials measuring 21 × 70 mm (15-cm$^3$ capacity) are excellent. Each vial can be poured with about 10 cm$^3$ medium, and 10 vials fit conveniently in a Pyrex storage jar measuring 80 × 100 mm. The tubes must have a flat bottom so they will stand unsupported. Each tube is capped individually with aluminum foil and then placed in the storage jar. The entire unit is then wrapped in aluminum foil. Also, the Petri dishes are wrapped with heavy-duty aluminum foil.

8. Weigh 0.2 g Gelrite and 3.0 g reagent-grade sucrose, and transfer them to a 250-cm$^3$ Erlenmeyer flask. Add 100 cm$^3$ of the MS medium (step 6). Seal the flask with an aluminum foil cap and sterilize the medium with wet heat (Chapter 3).

9. While the medium is in the autoclave, clean the interior of the hood with a tissue soaked in 70% (v/v) ethanol. Arrange the sterile Petri dishes or culture tubes to receive the autoclaved medium.

10. After the sterilized medium is removed from the autoclave, the flasks are swirled for a few minutes to ensure the dissolution of the sucrose and to mix the Gelrite with the remainder of the medium prior to pouring into the culture tubes. The Erlenmeyer flasks that contained the medium should be washed immediately after use, that is, before the residual gel has solidified.

11. After the gel in the tubes has cooled, replace them in the storage jar and wrap the jar in aluminum foil. Store the units in the refrigerator until 1 hr before culture time.

## Note on the importance of pH

The pH of the nutrient medium is highly important since it influences the uptake of various components of the medium as well as regulating a wide range of biochemical reactions occurring in plant tissue cultures (Owen, Wengerd, & Miller, 1991). Although media are usually adjusted to pH 5.2–5.8 with NaOH and HCl before autoclaving, most media are poorly buffered, and the pH drifts during the course of an experiment (Martin, 1980). This fluctuation is due to the unequal uptake of cations and anions, that is, in exchange for H$^+$ and OH$^-$ across the plasma membrane. Numerous workers have reported that the high temperatures of autoclaving causes the pH to be altered from the preset value (Skirvin et al., 1986). Postautoclave pH may be influenced by the type of carbohydrate employed, the brand of the gelling agent, and the presence and type of activated charcoal (Owen et al., 1991). Plant tissue culture workers should be aware of postautoclave changes in pH, as well as drifts in pH that will occur during an experiment. In reporting results, authors should state whether the pH was adjusted before or after autoclaving (Skirvin et al., 1986). The development of nutrient media with increased buffering capacity is clearly needed in order to prevent these changes in pH. A buffer, MES (2-[N-morpholino]ethanesulfonic acid), is available from Sigma and has been used in plant tissue culture media (Parfitt, Almehdi, & Bloksberg, 1988).

### SELECTED REFERENCES

Anderson, W. C. (1980). Tissue culture propagation of red and black raspberries, *Rubus idaeus* and *R. occidentalis*. *Acta Hortic.* 112, 13–20.
Ball, E. (1953). Hydrolysis of sucrose by autoclaving media, a neglected aspect in the technique of culture of plant tissues. *Bull. Torrey Bot. Club* 80, 409–11.
Bandurski, R. S. (1984). Metabolism of indole-3-acetic acid. In *The biosynthesis and metabolism of plant hormones*, ed. A. Crozier & J. R. Hillman, pp. 183–200. Cambridge: Cambridge University Press.
Bonga, J. M. (1982). Tissue culture techniques. In *Tissue culture in forestry*, ed. J. M. Bonga & D. J. Durzan, pp. 4–35. The Hague: Martinus Nijhoff/Junk.
Bragt, J. van, & Pierik, R. L. M. (1971). The effect of autoclaving on the gibberellin activity of aqueous solutions containing gibberellin $A_3$. In *Effects of sterilization on components in nutrient media*, ed. J. van Bragt, D. A. A. Mossel, R. L. M. Pierik, & H. Veldstra, pp. 133–7. Wageningen: H. Veenman & Zonen.
Brown, D. C. W., Leung, D. W. M., & Thorpe, T. A. (1979). Osmotic requirement for shoot formation in tobacco callus. *Physiol. Plant.* 46, 36–41.
Brown, D. C. W., & Thorpe, T. A. (1980). Changes in water potential and its components during shoot formation in tobacco callus. *Physiol. Plant.* 49, 83–7.
Brown, S., Wetherell, D. F., & Dougall, D. K. (1976). The potassium requirement for growth and embryogenesis in wild carrot suspension cultures. *Physiol. Plant.* 37, 73–9.
Cheng, T. Y., & Voqui, T. H. (1977). Regeneration of Douglas fir plantlets through tissue culture. *Science* 198, 306–7.
Chu, C-C. (1978). The $N_6$ medium and its application to anther culture of cereal crops. In *Proceedings of the symposium on plant tissue culture*, pp. 43–50. Peking: Science Press.

Chu, C-C., Wang, C-C., Sun, C-S., Hsu, C., Yin, K-C., Chu, C-Y., & Bi, F-Y. (1975). Establishment of an efficient medium for anther culture of rice through comparative experiments on the nitrogen sources. *Scientia Sinica 18*, 659–68.

Collins, G. B., Vian, W. E., & Phillips, G. C. (1978). Use of 4-amino-3,4,6-trichloropicolinic acid as an auxin source in plant tissue cultures. *Crop Science 18*, 286–8.

Constantin, M. J., Henke, R. R., & Mansur, M. A. (1977). Effect of activated charcoal on callus growth and shoot organogenesis in tobacco. *In Vitro 13*, 293–6.

Dekhuijzen, H. M. (1971). Sterilization of cytokinins. In *Effects of sterilization on components in nutrient media*, ed. J. van Bragt, D. A. A. Mossel, R. L. M. Pierik, & H. Veldstra, pp. 129–32. Wageningen: H. Veenman & Zonen.

Delmer, D. P. (1979). Dimethylsulfoxide as a potential tool for analysis of compartmentation in living plant cells. *Plant Physiol. 64*, 623–9.

Digby, J., & Skoog, F. (1966). Cytokinin activation of thiamine biosynthesis in tobacco callus cultures. *Plant Physiol. 41*, 647–52.

Dixon, R. A. (1985). Isolation and maintenance of callus and cell suspension cultures. In *Plant cell culture: A practical approach*, ed. R. A. Dixon, pp. 1–20. Oxford: IRL Press.

Doley, W. P., & Saunders, J. W. (1989). Hormone-free medium will support callus production and subsequent shoot regeneration from whole leaf explants in some sugarbeet (*Beta vulgaris* L.) populations. *Plant Cell Reports 8*, 222–5.

Dravnicks, D. E., Skoog, F., & Burris, R. H. (1969). Cytokinin activation of de novo thiamine biosynthesis in tobacco callus cultures. *Plant Physiol. 44*, 866–70.

Dunlap, J. R., Kresovich, S., & McGee, R. E. (1986). The effect of salt concentration on auxin stability in culture media. *Plant Physiol. 81*, 934–6.

Einset, J. W. (1978). Citrus tissue culture: Stimulation of fruit explant cultures with orange juice. *Plant Physiol. 62*, 885–8.

Eriksson, T. (1965). Studies on the growth requirements and growth measurements of cell cultures of *Haplopappus gracilis*. *Physiol. Plant. 18*, 976–93.

Favero, M. S., Carson, L. A., Bond, W. W., & Petersen, N. J. (1971). *Pseudomonas aeurginosa*: Growth in distilled water from hospitals. *Science 173*, 836–8.

Fridborg, G., Petersen, M., Landstrom, L. E., & Eriksson, T. (1978). The effect of activated charcoal on tissue cultures: Absorption of metabolites inhibiting morphogenesis. *Physiol. Plant. 43*, 104–6.

Gamborg, O. L., Miller, R. A., & Ojima, K. (1968). Nutrient requirements of suspension cultures of soybean root cells. *Exp. Cell Res. 50*, 151–8.

Gamborg, O. L., Murashige, T., Thorpe, T. A., & Vasil, I. K. (1976). Plant tissue culture media. *In Vitro 12*, 473–8.

Gamborg, O. L., & Shyluk, J. P. (1981). Nutrition, media and characteristics of plant cell and tissue cultures. In *Plant tissue culture: Methods and applications in agriculture*, ed. T. A. Thorpe, pp. 21–44. New York: Academic Press.

Hagen, S. R., Muneta, P., Augustin, J., & LeTourneau, D. (1991). Stability and utilization of picloram, vitamins, and sucrose in a tissue culture medium. *Plant Cell, Tissue & Organ Cult. 25*, 45–8.

Hangarter, R. P., & Stasinopoulos, T. C. (1991). Effect of Fe-catalyzed photooxidation of EDTA on root growth in plant culture media. *Plant Physiol. 96*, 843–7.

Heller, R. (1965). Some aspects of the inorganic nutrition of plant tissue cultures. In *Proceedings of the international conferences on plant tissue culture*, ed. P. R. White & A. R. Grove, pp. 1–17. Berkeley: McCutchan.

Horsch, R. B., King, J., & Jones, G. E. (1980). Measurement of cultured plant cell growth on filter paper discs. *Can. J. Bot. 58*, 2402–6.

Huang, L. C., & Murashige, T. (1977). Plant tissue culture media: Major constituents, their preparation and some applications. *Tissue Culture Assoc. Manual 3*, 539–48.

Klein, R. M., & Klein, D. T. (1970). *Research methods in plant science*. Garden City, N.Y.: Natural History Press.

Klucas, R. V., Hanus, F. J., & Russell, S. A. (1983). Nickel: A micronutrient element for hydrogen-dependent growth of *Rhizobium japonicum* and for the expression of urease activity in soybean leaves. *Proc. Nat. Acad. Sci. USA 90*, 2253–7.

Kohlenbach, H. W., & Wernicke, W. (1978). Investigations on inhibitory effect of agar and function of active carbon in anther culture. *Z. Pflanzenphysiol. 86*, 463–72.

Levitt, J. (1974). *Introduction to plant physiology*, 2d Ed. Saint Louis: Mosby.

Linsmaier-Bednar, S. M., & Skoog, F. (1967). Thiamine requirement in relationship to cytokinin in "normal" and "mutant" strains of tobacco callus. *Planta 72*, 146–54.

Lloyd, G., & McCown, B. (1980). Commercially-feasible micropropagation of mountain laurel, *Kalmia latifolia*, by use of shoot-tip culture. *Int. Plant Propagation Soc. Proc. 30*, 421–7.

Marschner, H. (1986). *Mineral nutrition of higher plants*. Orlando, Fla.: Academic Press.

Martin, S. M. (1980). Environmental factors. B. Temperature, aeration, and pH. In *Plant tissue culture as a source of biochemicals*, ed. E. J. Staba, pp. 143–8. Boca Raton: CRC Press.

Mattson, J. S., & Mark, J. B., Jr. (1971). *Activated carbon*. New York: Marcel Dekker.

Meins, F., Jr., & Binns, A. N. (1982). Rapid reversion of cell-division factor habituated cells in cultures. *Differentiation 23*, 10–12.

Meins, F., Jr., & Lutz, J. (1980). The induction of cytokinin habituation in primary pith explants of tobacco. *Planta 149*, 402–7.

Miller, A. R., & Roberts, L. W. (1984). Ethylene biosynthesis and xylogenesis in *Lactuca* pith explants cultured in vitro in the presence of auxin and cytokinin. The effect of ethylene precursors and inhibitors. *J. Exp. Bot. 35*, 691–8.

Miller, L. R., & Murashige, T. (1976). Tissue culture propagation of tropical foliage plants. *In Vitro 12*, 797–813.

Morel, G., & Muller, J. R. (1964). La culture in vitro du méristème apical de la pomme de terre. *C. R. Acad. Sci. (Paris) 258*, 5250–2.

Morré, D. J., Boss, W. F., & Loewus, F. A. (1990). *Inositol metabolism in plants*. New York: Wiley–Liss.

Murashige, T., & Skoog, F. (1962). A revised medium for rapid growth and bioassays with tobacco tissue cultures. *Physiol. Plant. 15*, 473–97.

Nissen, S. J. & Sutter, E. G. (1990). Stability of IAA and IBA in nutrient medium to several tissue culture procedures. *HortScience 25*, 800–2.

Nitsch, J. P., & Nitsch, C. (1955). Action synergique des auxines et du jus de tomate sur la croissance de tissus végétaux cultivés in vitro. *Soc. Bot. France Bull. 102*, 519–27.

(1969). Haploid plants from pollen grains. *Science 163*, 85–7.

Ohira, K., Ikeda, M., & Ojima, K. (1976). Thiamine requirements of various plant cells in suspension culture. *Plant & Cell Physiol. 17*, 583–8.

Owen, H. R., & Miller, A. R. (1992). An examination and correction of plant tissue culture basal medium formulations. *Plant Cell, Tissue & Organ Cult. 28*, 147–50.

Owen, H. R., Wengerd, D., & Miller, A. R. (1991). Culture medium pH is influenced by basal medium, carbohydrate source, gelling agent, activated charcoal, and medium storage method. *Plant Cell Reports 10*, 583–6.

Parfitt, D. E., Almehdi, A. A., & Bloksberg, L. N. (1988). Use of organic buffers in plant tissue culture systems. *Sci. Hortic. 36*, 157–63.

Peer, H. G. (1971). Degradation of sugars and their reactions with amino acids. In *Effects of sterilization on components in nutrient media*, ed. J. van Bragt, D. A. A. Mossel, R. L. M. Pierik, & H. Veldstra, pp. 105–15. Wageningen: H. Veenman & Zonen.

Phillips, R., & Dodds, J. H. (1977). Rapid differentiation of tracheary elements in cultured explants of Jerusalem artichoke. *Planta 135*, 207–12.

Pierik, R. L. M. (1971). Plant tissue culture as motivation for the symposium. In *Effects of sterilization on components in nutrient media*, ed. J. van Bragt, D. A. A. Mossel, R. L. M. Pierik, & H. Veldstra, pp. 3–13. Wageningen: H. Veenman & Zonen.

(1976). *Anthurium andraeanum* plantlets produced from callus tissues cultivated *in vitro*. *Physiol. Plant. 37*, 80–2.

Poovaiah, B. W., Reddy, A. S. N., & McFadden, J. J. (1987). Calcium messenger system – role of protein phosphorylation and inositol bisphospholipids. *Physiol. Plant. 69*, 569–73.
Proskauer, K., & Berman, R. (1970). Agar culture medium modified to approximate soil conditions. *Nature 227*, 1161.
Rédei, G. P. (1974). 'Fructose effect' in higher plants. *Ann. Bot. (Lond.) 38*, 287–97.
Riopel, J. L. (1973). *Experiments in developmental botany*. Dubuque, Iowa: Brown.
Robbins, W. J., & Hervey, A. (1974). Toxicity of water stored in polyethylene bottles. *Bull. Torrey Bot. Club 101*, 287–91.
Roberts, L. W. (1988). Hormonal aspects of vascular differentiation. In *Vascular differentiation and plant growth regulators*, ed. L. W. Roberts, P. B. Gahan, & R. Aloni, pp. 22–38. Heidelberg: Springer–Verlag.
Roberts, L. W., & Baba, S. (1978). Exogenous methionine as a nutrient supplement for the induction of xylogenesis in lettuce pith explants. *Ann. Bot. (Lond.) 42*, 375–9.
Roblin, G., & Fleurat-Lessard, P. (1983). Dimethylsulfoxide action on dark-induced and light-induced leaflet movements and its necrotic effects on excised leaves of *Cassia fasiculata. Physiol. Plant. 58*, 493–6.
Schenk, R. U., & Hildebrandt, A. C. (1972). Medium and techniques for induction and growth of monocotyledonous and dicotyledonous plant cell cultures. *Can. J. Bot. 50*, 199–204.
Schmitz, R. Y., & Skoog, F. (1970). The use of dimethylsulfoxide as a solvent in the tobacco bioassay for cytokinins. *Plant Physiol. 45*, 537–8.
Schneider, V. F., Emmerich, A., & Akyar, O. C. (1975). Occlusion of nonsucrose substances during crystallization of sucrose. *Zucker 28*, 113–21.
Skirvin, R. M., Chu, M. C., Mann, M. L., Young, H., Sullivan, J., & Fermanian, T. (1986). Stability of tissue culture medium pH as a function of autoclaving, time, and cultured plant material. *Plant Cell Reports 5*, 292–4.
Smith, R. H. (1992). *Plant tissue culture: Techniques and experiments*. San Diego: Academic Press, Inc.
Stasinopoulos, T. C., & Hangarter, R. P. (1990). Preventing photochemistry in culture media by long-pass light filters alters growth of cultured tissues. *Plant Physiol. 93*, 1365–9.
Stehsel, M. L., & Caplin, S. M. (1969). Sugars: Autoclaving vs. sterile filtration on the growth of carrot root tissue in culture. *Life Sci. 8*, 1255–9.
Steiner, A. A., & van Winden, H. (1970). Recipe for ferric salts of ethylene diaminetetraacetic acid. *Plant Physiol. 46*, 862–3.
Straus, J. (1960). Maize endosperm tissue growth in vitro. III. Development of a synthetic medium. *Am. J. Bot. 47*, 641–7.

Street, H. E. (1966). The nutrition and metabolism of plant tissue and organ cultures. In *Cells and tissue in culture*, ed. E. N. Willmer, vol. 3, pp. 533–630. New York: Academic Press.
  (1969). Growth in organized and unorganized systems. In *Plant physiology: A treatise*, ed. F. C. Steward, vol. VB, pp. 3–224. New York: Academic Press.
  (ed.) (1973). *Plant tissue and cell cultures*. Oxford: Blackwell Scientific Publications.
Tabor, C. A. (1981). Improving the suitability of glass fiber filters for use as culture supports. *In Vitro 17*, 129–32.
Ten Ham, E. J. (1971). Vitamins. In *Effects of sterilization on components in nutrient media*, ed. J. van Bragt, D. A. A. Mossel, R. L. M. Pierik, & H. Veldstra, pp. 121–3. Wageningen: H. Veenman & Zonen.
Thorpe, T. A. (1982). Physiological and biochemical aspects of organogenesis in vitro. In *Plant tissue culture 1982*, ed. A. Fujiwara, pp. 121–4. Tokyo: Japanese Association for Plant Tissue Culture.
Vannini, G. L., & Poli, F. (1983). Binucleation and abnormal chromosome distribution in *Euglena gracilis* cells treated with dimethylsulfoxide. *Protoplasma 114*, 62–6.
Veith, R., & Komor, E. (1991). Nutrient requirement for optimal growth of sugarcane suspension cells: Nicotinic acid is an essential growth factor. *J. Plant Physiol. 139*, 175.
Wang, P. J., & Huang, L. C. (1976). Beneficial effects of activated charcoal on plant tissue and organ culture. *In Vitro 12*, 260–2.
Warren Wilson, J., Roberts, L. W., Warren Wilson, P. M., & Gresshoff, P. M. (1994). Stimulatory and inhibitory effects of sucrose concentration on xylogenesis in lettuce pith explants: Possible mediation by ethylene biosynthesis. *Ann. Bot. (Lond.) 73*, 65–73.
Weatherhead, M. A., Burdon, J., & Henshaw, G. G. (1978). Some effects of activated charcoal as an additive to plant tissue culture media. *Z. Pflanzenphysiol. 89*, 141–7.
White, P. R. (1934). Potentially unlimited growth of excised tomato root tips in a liquid medium. *Plant Physiol. 9*, 585–600.
  (1943). *A handbook of plant tissue culture*. Lancaster, Pa.: Jacques Cattell Press.
  (1963). *The cultivation of animal and plant cells*. New York: Ronald Press Company.
Wilmar, J. C., & Doornbos, T. (1971). Stability of abscisic acid and isomers to heat sterilization and light. In *Effects of sterilization on components in nutrient media*, ed. J. van Bragt, D. A. A. Mossel, R. L. M. Pierik, & H. Veldstra, pp. 139–47. Wageningen: H. Veenman & Zonen.
Windholz, M. (ed.) (1983). *The Merck index: An encyclopedia of chemicals, drugs, and biologicals*, 10th Ed. Rahway, N.J.: Merck & Co., Inc.
Wolter, K. E., & Skoog, F. (1966). Nutritional requirements of *Fraxinus* callus cultures. *Am. J. Bot. 53*, 263–9.

Wright, K., & Northcote, D. H. (1972). Induced root differentiation in sycamore callus. *J. Cell Sci. 11*, 319–77.

Yeoman, M. M. (1973). Tissue (callus) cultures – techniques. In *Plant tissue and cell culture*, ed. H. E. Street, pp. 31–58. Oxford: Blackwell Scientific Publications.

PART II

*Experimental: Callus and callus-derived systems*

# 5

# Initiation and maintenance of callus

A callus consists of an amorphous mass of loosely arranged thin-walled parenchyma cells arising from the proliferating cells of the cultured explant. Frequently, as a result of wounding, a callus is formed at the cut end of a stem or root. The term "callus" should not be confused with "callose," another botanical term. The latter refers to a polysaccharide associated primarily with sieve elements (Esau, 1977). Although the major emphasis has been on angiosperm tissues, callus has been observed in gymnosperms, ferns, mosses, and liverworts (Yeoman, 1970; Yeoman & Macleod, 1977).

Sinnott (1960) has described some of the early observations on wound callus formation. The stimuli involved in the initiation of wound callus are the endogenous hormones auxin and cytokinin. In addition to mechanical injury, callus may be produced in plant tissues following an invasion by certain microorganisms (Braun, 1954) or by insect feeding (Pelet et al., 1960). Using tissue culture techniques, callus formation can be induced in numerous plant tissues and organs that do not usually develop callus in response to an injury (Street, 1969). Plant material typically cultured includes vascular cambia, storage parenchyma, pericycle of roots, cotyledons, leaf mesophyll, and provascular tissue. In fact, all multicellular plants are potential sources of explants for callus initiation (Yeoman & Macleod, 1977).

In 1939 the first successful prolonged cultures of experimentally induced callus were achieved almost simultaneously at the research laboratories of Gautheret in Paris, Nobécourt in Grenoble, and White in Princeton. These cultures were originally derived from explants of cambial tissue of carrot and tobacco. The term "tissue culture," as applied to such cultures, is a misnomer: A cultured tissue does not maintain its unique characteristics as a plant tissue, but reverts to a disorganized callus. The most important characteristics of callus, from a functional viewpoint, is that it has the potential to develop normal roots, shoots, and embryoids that can form plants and, in addition, can be used to initiate a suspension culture.

Establishment of a callus from an explant can be divided roughly into three developmental stages: induction, cell division, and differentiation. During the initial induction phase metabolism is stimulated prior to mitotic activity. The length of this phase depends on the physiological status of the explant cells as well as the cultural conditions. Subsequently, there is a phase of active cell division as the explant cells revert to a meristematic state. The third phase involves the appearance of cellular differentiation and the expression of certain metabolic pathways that lead to the formation of secondary products. Secondary product biosynthesis in relation to callus differentiation has been reviewed by Yeoman and his colleagues (1982).

The growth characteristics of a callus involve a complex relationship among the plant material used to initiate the callus, the composition of the medium, and the environmental conditions during the incubation period (Aitchison, Macleod, & Yeoman, 1977). Some callus growths are heavily lignified and hard in texture, whereas others break easily into small fragments. Fragile growths that crumble readily are termed "friable cultures." Callus may appear yellowish, white, green, or pigmented with anthocyanin. Pigmentation may be uniform throughout the callus or some regions may remain unpigmented. Anthocyanin-synthesizing and -nonsynthesizing cell lines have been isolated from carrot cultures (Alfermann & Reinhard, 1971), and a stable pigment-producing strain of cultured *Euphorbia* sp. cells was isolated after 24 clonal selections and subcultures (Yamamoto, Mizuguchi, & Yamada, 1982).

There is considerable variability in the anatomy of callus cultures. A homogeneous callus consisting entirely of parenchyma cells is rarely found, although exceptions have been reported for *Agave* and *Rosa* cultures (Narayanaswamy, 1977). Cytodifferentiation occurs in the form of tracheary elements, sieve elements, suberized cells, secretory cells, and trichomes. Small nests of dividing cells form vascular nodules (meristemoids) that may become centers for the formation of shoot apices, root primordia, or incipient embryos. Vascular nodules typically consist of discrete zones of xylem and phloem separated by a cambium. The orientation of the xylem and phloem with respect to the cambial zone is influenced by the nature of the original tissue (Gautheret, 1959, 1966). The location of the nodules within the callus can be modified by altering the composition of the medium (Chapter 13). Vascular differentiation may also take the form of somewhat randomly arranged strands of tracheary elements (Roberts, 1988).

The hormonal requirements for the initiation of callus depend on the origin of the explant tissue. Juice vesicles from lemon fruits (Kordan, 1959), and explants containing cambial cells, exhibit callus growth without the addition of any exogenous growth regulators. Most excised tissues, however, require the addition of one or more growth regulators in order to initiate callus formation (Yeoman & Macleod, 1977). Explants can be classified according to their exogenous requirements, in the following manner: (1) auxin, (2) cytokinin, (3) auxin and cytokinin, and (4) complex natural extracts. Supplements for callus initiation are given in Chapter 4.

After the callus has been grown for a while in association with the original tissue, it becomes necessary to subculture the callus to a fresh medium. Growth on the same medium for an extended period will lead to a depletion of essential nutrients and to a gradual desiccation of the gelling agent. Metabolites secreted by the growing callus may accumulate to toxic levels in the medium. The transferred fragment of callus must be of a sufficient size to ensure renewed growth on the fresh medium. If the transferred inoculum is too small, it may exhibit a very slow rate of growth or none at all. Street (1969) recommended that the inoculum be 5–10 mm in diameter and weigh 20–100 mg. Successive subcultures are usually performed every four to six weeks with cultures maintained on an agar medium at 25 °C or above (Yeoman & Macleod, 1977). Passage time, however, is somewhat variable and depends on the rate of growth of the callus. A typical growth curve for callus cultures is shown in Figure 5.1; it resembles growth curves plotted for bacterial cell cultures. A friable callus can be subdivided with a thin spatula or scalpel, transferred directly to the surface of a sterile Petri dish, and sliced into fragments with a scalpel. Only healthy tissue should be transferred, and brown or necrotic tissue must be discarded. Interest has been shown in developing alternative methods for long-term maintenance of tissue cultures, for example, freeze preservation (Withers, 1979; see Chapter 16).

What are some of the best plant materials for the initiation of a callus culture? Young healthy tissues that are rich in nutrients, and possibly endogenous hormones, are the best choices for the induction of cell division; for example, storage organs and cotyledons of seeds. These include tissues from potato tuber (*Solanum tuberosum*), storage roots of turnip (*Brassica rapa*), sweet potato (*Ipomea batatas*), and carrot (*Daucus carota*). Also, callus is easily started from the cotyledons of soybean (*Glycine max*). Stem pith parenchyma from lettuce (*Lactuca sativa*) and tobacco (*Nicotiana tabacum*) readily divides in the presence of auxin and cytokinin.

Figure 5.1. Growth response of a typical callus culture. This particular callus should be subcultured approximately at the time indicated by X.

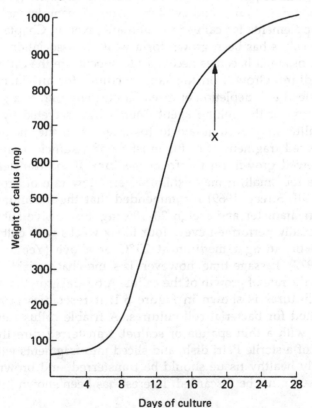

Another approach is to germinate seeds under aseptic conditions and subsequently take explants from the axenic seedlings. For very small seeds, such as radish (*Raphanus sativus*), it is recommended that the seeds be carried through the sterilization procedure in a small porous bag. One simple device is an empty paper tea bag, which, of course, usually has a string attached. Woody plant material is generally a poor choice. Also, plant tissues that are high in oxidase activity pose a special problem since enzymatic browning retards cell division. The browning results from the activity of wound-induced copper oxidases (polyphenoloxidase). This may be suppressed, to some extent, by the use of an antioxidant mixture (Sigma).

The purpose of the following experiment is to acquaint the student with the technique of inducing callus formation in explants excised from a potato tuber (*Solanum tuberosum* L.). Chapman (1955) was the first to initiate callus from a potato tuber on a medium supplemented with coconut water and 2,4-D. Although other workers have subsequently grown potato callus on various concentrations of kinetin and 2,4-D (Shaw et al., 1976; LaRosa, Hasegawa, & Bressan, 1984; Van der Plas & Wagner, 1984), a common problem has been sluggish growth. Hagen and his colleagues (1990) found that the synthetic auxin picloram was highly effective in initiating and maintaining potato callus. Picloram, the active ingredient in the herbicide Tordon, was effective as the only growth regulator in the medium of Hagen's group. For most potato cultivars a concentration of 10 m$M$ (2.41 mg/l) was optimal, although one cultivar responded better to 20 m$M$. The medium employed in the present experiment will be supplemented with 3.0 mg/l picloram (Chapter 4).

### LIST OF MATERIALS

*Sterilization mode:* C, chemical; O, oven; A, autoclave

- C  one or more healthy potato (*Solanum tuberosum* L.) tubers
- –  paring knife; vegetable scraper
- –  1,000 cm$^3$ aqueous solution (20% v/v) commercial bleach containing a final concentration of approximately 1.0% (v/v) NaOCl (tuber pretreatment)
- –  300 cm$^3$ aqueous solution (10% v/v) bleach (tissue block sterilization)

- O  600-cm³ beakers (four)
- O  9-cm³ Petri dishes (five)
- O  stainless steel cork borer (No. 4; 8.5–9.0 mm I.D.) containing a metal rod; enclose borer in a test tube capped with aluminum foil, and place an additional layer of foil around the entire unit
- O  stainless steel forceps; enclose in a test tube and wrap with foil in a manner similar to that of the cork borer
- O  culture tubes, 21 × 70 mm (15-cm³ capacity) recommended (three); cap tubes with foil, place in storage jars (80 × 100 mm), and wrap the units with foil
- A  125-cm³ Erlenmeyer flasks, each containing 100 cm³ $DDH_2O$ (12)
- A  9-cm Petri dishes, each containing two sheets Whatman No. 1 filter paper (five); enclose the dishes in a paper bag
- A  300 cm³ MS callus-induction medium (Chapter 4) supplemented with sucrose (3.0% w/v) and Gelrite (0.2% w/v); prepare 3 × 100 cm³ aliquots in 250-cm³ Erlenmeyer flasks
- A  9-cm³ Petri dish equipped as explant cutting guide (directions in Appendix); enclose the device in a paper bag
- C  scalpel; Bard–Parker No. 7 surgical knife handle with a No. 10 blade is recommended
- –  ethanol (80% v/v) dip for flaming instruments; place the dip in a metal can
- –  ethanol (70% v/v) in a plastic squeeze bottle for surface disinfection of hood
- –  methanol lamp
- –  interval timer
- –  heavy-duty aluminum foil (roll)
- –  light microscope (100× magnification)
- –  dissecting microscope or hand lens
- –  dissecting needles
- –  microscope slides
- –  cover slips
- –  lens paper
- –  aqueous solution toluidine blue O (0.05% w/v) in dropper bottle
- –  Pasteur pipette
- –  Tween 20
- –  surgical gloves

## PROCEDURE

Follow the instructions in the Appendix to Chapter 4 for the preparation of 1 liter of MS medium supplemented with picloram (3.0 mg/l), sucrose (3.0% w/v), and Gelrite (0.2% w/v). Prepare to culture 30 potato tuber explants, which will require the preparation of 30 culture tubes of medium (10 cm$^3$ each).

If the experiment is conducted as a classroom project, each student may be assigned to a different cultivar of potato. Another possible assignment is to examine the relative effectiveness of different concentrations of picloram on the induction of callus growth.

1. Directions for preparing a tuber for explant removal were given in Chapter 3. Wash the tuber with water containing a detergent, followed by running tap water. The tuber is given a pretreatment by immersion in a liter of 20% bleach solution containing Tween 20 (10 drops) for 30 min. It is advisable to wear surgical gloves, and care must be taken not to splash any of the hypochlorite solution on the skin or clothing. Rinse the tubers briefly with tap water. With a paring knife and vegetable scraper remove the outer 1–2 mm of suberized periderm and carefully remove all traces of buds ("eyes") and surface discoloration. Each student will need a minimum of six blocks of tissue, so it would be best to prepare one or two additional tissue blocks. All subsequent steps must be conducted under aseptic conditions.

2. Disinfect the working area within the hood with a tissue soaked with ethanol (70% v/v). Place the tissue blocks in one of the sterile 600-cm$^3$ beakers. Add the 10% (v/v) hypochlorite solution to the beaker, and set the timer for 10 min. If the hood is equipped with a UV lamp, *do not* turn on the lamp; otherwise, UV-induced degradation of the hypochlorite solution might occur (Chapter 3).

3. After about 8 min, preparations can be started for rinsing the tissue blocks. Thoroughly wash your hands with soap and hot water before starting the aseptic procedure. Unwrap the remaining three beakers (600 cm$^3$), and add about 300 cm$^3$ sterile DDH$_2$O to each beaker. Following the 10-min sterilization period, remove the tissue blocks from the hypochlorite solution with the forceps, and rinse them successively for 30 sec to 1 min in each of the three rinse beakers. Withdraw from the hood the beakers containing the hypochlorite solution and the first two rinses. Light the

methanol lamp and place the scalpel in the ethanol dip. Open the foil packet containing the cork borer and the packet of empty Petri dishes. Place two of the dishes in the rear of the hood, fill each dish about halfway with sterile $DDH_2O$, and partially remove the lid of one of the dishes. Remove the lid of the third empty Petri dish, exposing the sterile inner surface of the lower half (boring platform).

4. Flame the forceps and transfer one of the tissue blocks from the final rinse beaker to the boring platform. Steady the tissue with the forceps, and make a single vertical boring with the cork borer through the center of the tissue block. The cork borer must be inserted all the way so that it cuts completely through the tissue block. Lift the block with the borer still inserted in it, and hold the block directly over the $DDH_2O$ in the partially opened Petri dish. Gently exert pressure on the metal rod. This slight force should eject the tissue cylinder into the pool of water. Return the tissue block to the boring platform. Place the arms of the forceps on each side of the borer and withdraw the borer from the tissue. Repeat the process with the other blocks of tissue until you have prepared six or seven tissue cylinders. Finally, withdraw from the hood the final rinse beaker, cork borer, remains of the potato tissue, and the boring platform.

5. Open the packet containing the explant cutting guide (Appendix). If this device is unavailable, the bottom half of a sterile Petri dish can be used for slicing explants from the cylindrical tissue borings. Arrange the following three Petri dishes in the rear of the hood. The nearest dish should be the explant cutting guide, another dish will contain the cylinders of tissue, and the third dish contains sterile $DDH_2O$ for explant rinsing (prepared in step 3). Partially open the two plates in the rear of the hood and completely remove the lid of the explant cutting guide.

6. Flame the forceps and scalpel, and transfer a tissue cylinder with the forceps to the cutting guide. The flamed instruments should be permitted to cool briefly before bringing them in contact with living plant tissues. Trim and discard approximately 2 mm of tissue from each end of the cylinder. Slice the remaining cylinder into segments of 2–3 mm in length. Each explant, therefore, will measure about 8 mm in diameter and 2–3 mm in thickness. Each cylinder should yield a minimum of 5 explants. With the flat blade of the scalpel, transfer the explants to the plate con-

taining the DDH$_2$O rinse. Repeat the cutting operation until 30 explants have been prepared. Flame the forceps and scalpel several times during the course of the slicing operation. Remove from the hood the Petri plate that contained the cylinders of tissue and the cutting guide.

7. Fill an empty Petri dish halfway with DDH$_2$O. Flame the forceps and transfer the explants to the plate containing the rinse water. Remove from the hood the plate that formerly contained the explants. Open the paper bag containing the Petri dishes with the Whatman No. 1 filter paper. Arrange the culture tubes to receive the explants.

8. Flame the forceps and transfer the explants one at a time to the surface of the sterile filter paper. Blot briefly both the top and bottom of each explant; immediately transfer the explant to the surface of the culture medium (one explant per tube). Remember to hold the culture tubes at a slight angle so that the hand grasping the forceps is not directly over the sterile surface of the medium.

9. After capping the culture tubes with foil, place them in the Pyrex storage jars and wrap the jars with foil. Transfer the cultured explants to an incubator adjusted to 25–27 °C.

## RESULTS

After a few days in culture, the explants become slightly rough in texture, and their surface may glisten in reflected light. This is a sign of the beginning of callus formation. Culture for a single incubation period (passage) may last from a few weeks to three months, depending on the rapidity of growth (Thomas & Davey, 1975). For most potato cultivars, within four to six weeks after initiating a culture with picloram, there should be more than 1 g fresh weight of callus available for subculture. Subcultured callus typically increases tenfold in fresh weight within four or five weeks on the MS–picloram medium solidified with Gelrite (Hagen et al., 1990). Depending on the friability of the callus, use either a spatula or a scalpel for transferring inocula from the callus mass to the fresh medium. Only gray or cream-colored tissue can be used; brownish tissue is a sign that localized necrosis has occurred. The instruments must be flamed and aseptic techniques used throughout the subculture procedure.

Examine the surface of the callus with a dissecting microscope or a hand lens, and notice the external appearance of the callus. With a dissecting needle scrape some of the cells onto a microscope slide. Add a drop of distilled water and a cover slip, and examine the cells with the light microscope (100× magnification). The contrast can be enhanced by lightly staining the cells with an aqueous solution of toluidine blue O (0.05% w/v; McCully & O'Brien, 1969). With a Pasteur pipette add a droplet of the stain solution to one edge of the cover slip. On the opposite side of the cover slip moisten a piece of lens paper with the aqueous mounting medium. The blotting action of the paper will draw the stain beneath the cover slip and into the field of vision.

## QUESTIONS FOR DISCUSSION

1. Why are explants containing cambial cells excellent choices for the initiation of callus cultures?
2. Under the cultural conditions employed in your experiment, what is the optimal interval of time between subcultures (passage time)? How is this time determined?
3. Give some reasons for subculturing.
4. What are vascular nodules?
5. What is the significance of each of the three stages of callus development?

## APPENDIX

*Initiation of callus from taproot of carrot (Daucus carota L.).* A culture of carrot callus should be initiated and maintained since it will be required in an experiment on somatic embryogenesis (Chapter 8). Obtain a large and healthy taproot of carrot fresh from the market, and remove the external 1–2 mm of tissue with a vegetable scraper. Cut the carrot transversely into 1-cm slices and surface sterilize them in the usual manner. Remove cylindrical tissue samples with a sterile cork borer (No. 3 or 4), and prepare explants in a procedure similar to that previously described for the potato tuber. The borings should be made in the vicinity of the vascular cambium for the most vigorous cellular proliferation (Gautheret, 1959). Culture the explants on an MS basal medium, MS vitamin supplement, *myo*-inositol (100 mg/l), sucrose (3.0% w/v), Gelrite (0.2% w/v), IAA (10 mg/l), and kinetin (0.1

mg/l). Transfer the cultures to an incubator (25–29 °C). The cultures must be maintained approximately four to six weeks before they are suitable for the induction of somatic embryogenesis.

*Cytokinin properties of coconut water.* Repeat the experiment as outlined in this chapter, except employ three different media. Each medium is prepared so as to contain a different combination of plant growth regulators:

1. 2,4-D (3.0 mg/l);
2. 2,4-D (3.0 mg/l) and kinetin (0.2 mg/l); and
3. 2,4-D (3.0 mg/l) and coconut water (10% v/v).

The procedure for the preparation of coconut water is given in the Appendix to Chapter 4. Can you detect any differences in the appearance of callus arising from the explants on the different media? Does this experiment provide any evidence for the presence of a cytokinin-like substance in coconut water?

*Explant cutting guide.* An aid for slicing explants can be constructed from a Petri dish, a glass rod, and an index card. Cut a small-diameter glass rod to a slightly smaller length than the inside diameter of the bottom half of the Petri dish, and cement the rod to the inside of the plate. On a 3 × 5-inch index card, draw two parallel lines about 2.5–3.0 mm apart. Trim the card and tape it to the outside of the bottom half of the plate with one of the lines parallel to the glass rod and directly beneath it. During explant preparation the cylinder of tissue is pressed against the rod and held perpendicular to it. The line on the card acts as a visual guide to indicate the position of the scalpel in cutting explants of approximately the same thickness. Wrap the device in a paper bag. Sterilize the cutting guide in the autoclave, and place it briefly in the drying oven to evaporate the condensed moisture from the autoclave.

*Cork borer.* The cutting edge of the cork borer can be maintained by occasionally rubbing it with a fine grade of abrasive paper. The Tufbak Durite T421 papers (400A, 500A) have given satisfactory results (Norton Consumer Products).

## SELECTED REFERENCES

Aitchison, P. A., Macleod, A. J., & Yeoman, M. M. (1977). Growth patterns in tissue (callus) cultures. In *Plant tissue and cell culture*, 2d Ed., ed. H. E. Street, pp. 267–306. Oxford: Blackwell Scientific Publications.

Alfermann, A. W., & Reinhard, E. (1971). Isolation of anthocyanin-producing and non-producing cell lines of tissue cultures of *Daucus carota*. *Experientia* 27, 353–4.

Braun, A. C. (1954). The physiology of plant tumors. *Annu. Rev. Plant Physiol.* 5, 133–62.

Chapman, H. W. (1955). Potato tissue cultures. *Am. Potato J.* 32, 207–10.

Esau, K. (1977). *Anatomy of seed plants*, 2d Ed. New York: Wiley.

Gautheret, R. J. (1959). *La culture des tissus végétaux: Techniques et réalisations*. Paris: Masson.

— (1966). Factors affecting differentiation of plant tissues grown *in vitro*. In *Cell differentiation and morphogenesis*, ed. W. Beermann, pp. 55–9. Amsterdam: North-Holland.

Hagen, S. R., LeTourneau, D., Muneta, P., & Brown, J. (1990). Initiation and culture of potato tuber callus tissue with picloram. *Plant Growth Regulation* 9, 341–5.

Kordan, H. A. (1959). Proliferation of excised juice vesicles of lemon in vitro. *Science* 129, 779.

LaRosa, P. C., Hasegawa, P. M., & Bressan, R. A. (1984). Photoautotropic potato cells: Transition from heterotropic to autotropic growth. *Physiol. Plant.* 61, 279–86.

McCully, M. E., & O'Brien, T. P. (1969). *Plant structure and development: A pictorial and physiological approach*. New York: Macmillan.

Narayanaswamy, S. (1977). Regeneration of plants from tissue cultures. In *Applied and fundamental aspects of plant cell, tissue and organ culture*, ed. J. Reinert & Y. P. S. Bajaj, pp. 179–206. Berlin: Springer–Verlag.

Pelet, F., Hildebrandt, A. C., Riker, A. J., & Skoog, F. (1960). Growth in vitro of tissues isolated from normal stems and insect galls. *Am. J. Bot.* 47, 186–95.

Roberts, L. W. (1988). Evidence from wound responses and tissue cultures. In *Vascular differentiation and plant growth & regulators*, ed. L. W. Roberts, P. B. Gahan, & R. Aloni, pp. 63–88. Heidelberg: Springer–Verlag.

Shaw, R., Varns, J. L., Miller, K. A., & Talley, E. A. (1976). Potato tuber callus. Validation as biochemical tool. *Plant Physiol.* 58, 464–7.

Sinnott, E. W. (1960). *Plant morphogenesis*. New York: McGraw–Hill.

Street, H. E. (1969). Growth in organized and unorganized systems. Knowledge gained by culture of organs and tissue explants. In *Plant physiology: A treatise*, vol. VB, ed. F. C. Steward, pp. 3–224. New York: Academic Press.

Thomas, E., & Davey, M. R. (1975). *From single cells to plants*. London: Wykeham.

Van der Plas, L. H. W., & Wagner, M. J. (1984). Influence of osmotic stress on the respiration of potato tuber callus. *Physiol. Plant. 62*, 398–403.
Withers, L. A. (1979). Freeze preservation of somatic embryos and clonal plantlets of carrot (*Daucus carota* L.). *Plant Physiol. 63*, 460–7.
Yamamoto, Y., Mizuguchi, R., & Yamada, Y. (1982). Selections of a high and stable pigment-producing strain in cultured *Euphorbia millii* cells. In *Plant tissue culture 1982*, ed. A. Fujiwara, pp. 283–4. Tokyo: Japanese Association for Plant Tissue Culture.
Yeoman, M. M. (1970). Early development in callus cultures. *Int. Rev. Cytol. 29*, 383–409.
Yeoman, M. M., Lindsey, K., Miedzybrodzka, M. B., & McLauchlan, W. R. (1982). Accumulation of secondary products as a facet of differentiation in plant cell and tissue cultures. In *Differentiation in vitro*, ed. M. M. Yeoman and D. E. S. Truman, pp. 65–82. Cambridge: Cambridge University Press.
Yeoman, M. M., & Macleod, A. J. (1977). Tissue (callus) cultures – techniques. In *Plant tissue and cell culture*, 2d Ed., ed. H. E. Street, pp. 31–59. Oxford: Blackwell Scientific Publications.

# 6

## Organogenesis

The capability to induce the formation of adventitious roots and shoots in vitro is of the utmost importance in plant tissue culture methodology. Studies involving the transformation of protoplasts would be of little value unless the genetically altered plant material could be regenerated into a plantlet. Plant regeneration by tissue culture techniques can be achieved by either zygotic embryo culture, somatic embryogenesis, or organogenesis. The latter approach is employed in micropropagation from bud and shoot material (Chapter 10) and in organ production from callus and suspension cultures. This chapter is devoted entirely to organogenesis. Roots, shoots, and flowers are the organs that may be initiated from tissue cultures. Embryos are not classified as organs because these structures have an independent existence; that is, embryos do not have vascular connections with the parent plant body.

The underlying basis for organogenesis is poorly understood and involves the interplay of a host of factors: donor plant growth, source of the explant, culture medium, supplements of growth regulators, and environmental conditions (Flick, Evans, & Sharp, 1983). The first major breakthrough came with the discovery that in vitro organogenesis in tobacco cultures could be chemically regulated. The addition of auxin to the medium served to initiate root formation, whereas shoot initiation was inhibited. The latter effect on shoot formation could be partially reversed by increasing the concentration of both sucrose and inorganic phosphate (Skoog, 1944). Later it was found that adenine sulfate was active in promoting shoot initiation, and this chemical reversed the inhibitory effect of auxin (Skoog & Tsui, 1948). The studies of Skoog's group led to the hypothesis that organogenesis is regulated by a balance between cytokinin and auxin. A relatively high auxin:cytokinin ratio induced root formation in tobacco callus, whereas a low ratio of the same compounds favored shoot production (Fig. 6.1; Skoog & Miller, 1957).

Probably the most precise regulation of organ formation has been achieved with epidermal and subepidermal explants consisting of a few cell layers in thickness (Tran Thanh Van, 1980a,b). The formation of floral buds, vegetative buds, and roots have been demonstrated in thin cell-layer explants of several species by regulating the auxin:cytokinin ratio, carbohydrate supply, and environmental conditions (Tran Thanh Van, Chlyah, & Chlyah,

Figure 6.1. The regulation of organ formation in explants of tobacco (*Nicotiana tabacum*) pith by varying the auxin:cytokinin ratio. Note the occurrence of shoots induced by high levels of kinetin. Moderate levels of kinetin, in the presence of auxin, stimulate the production of callus. Auxin induced root formation in the absence of kinetin. (Courtesy of F. Skoog.)

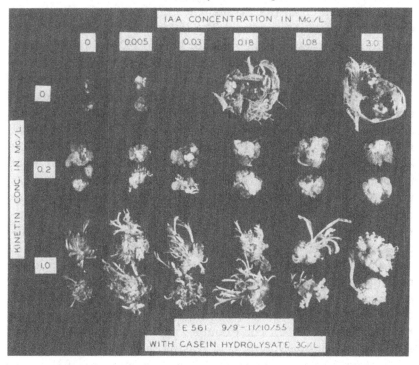

1974; Tran Thanh Van & Trinh, 1978). Certain isolated tissue layers in species that readily regenerate organs in vivo showed a remarkable potential to form organs during culture. Primary explants consisting of three to six cell layers of epidermal and subjacent collenchyma removed from the region of the leaf midvein of *Begonia rex* produced shoots or roots from the epidermal cells (Chlyah & Tran Thanh Van, 1975; Tran Thanh Van, 1980a). Root initiation occurred in the presence of $\alpha$-NAA plus zeatin, and shoot formation required the addition of either zeatin or benzylaminopurine in the absence of auxin. Organogenesis has been shown to occur in several other species from epidermal layers during culture (Tran Thanh Van & Trinh, 1978). Recent studies have shown that cell wall oligosaccharides influence organogenesis in thin cell layer explants of tobacco. The presence of plant growth regulators, pH, and the ionic environment are thought to be involved in the activation of certain hydrolytic cell wall enzymes that, in turn, release biologically active oligosaccharides (Tran Thanh Van & Mutaftschiev, 1990).

For plantlet regeneration in many dicot callus cultures, the callus is removed from the maintenance medium and subcultured on a shoot-induction medium. The latter medium usually has a cytokinin:auxin ratio in the range of 10:1 to 100:1, in many cases by supplementing the medium with cytokinin as the sole growth regulator. In comparison to dicots, monocot cultures are more difficult to regenerate. With monocot cultures exogenous cytokinin may be unnecessary for the initiation of shoots. The omission of auxin from the maintenance medium may suffice to induce shoot formation in these cultures, and two successive transfers on auxin-free media have been recommended (Gresshoff, 1978). Root initiation frequently occurs spontaneously after the culture has initiated buds, and shoot development undoubtedly alters the endogenous hormones within the culture (Gresshoff, 1978). Regenerated shoots are transferred to a root-inducing medium. In many cases, auxin alone or in combination with a low level of cytokinin will enhance root primordia formation. The appropriate cultural conditions for rhizogenesis in some species may be completely ineffective in a closely related plant (Street, 1977). There is some evidence that phenolic compounds may act with auxin to promote rooting (Thorpe, 1980). For example, the combination of phloroglucinol with indolebutyric acid was more effective in stimulating rooting than auxin alone (Welander & Huntrieser, 1981).

In addition to auxin and cytokinin, there are reports involving the possible roles of other growth regulators in the induction of organogenesis. Although there are a few exceptions (Flick et al., 1983), gibberellins tend to suppress both root and shoot initiation in cultures. Endogenous ethylene may be a factor in shoot initiation. One report indicates that ethylene blocks the early stages of organogenesis, but enhances the further development of primordia (Huxter, Reid, & Thorpe, 1979). Endogenous ethylene was identified as a factor in bud induction arising from cultured tobacco cotyledons (Everett, 1982). Indirect evidence suggests a similar role for ethylene in cultured *Lilium* bulb tissues (Aartrijk & Blom-Barnhoorn, 1983).

Cultured explants are typically incubated in the dark for the initiation and subsequent development of callus, although low-level illumination may be beneficial (Dixon, 1985). A light requirement has been reported for adventitious bud formation in hairy roots of horseradish (*Armoracia lapathifolia*). The roots had been inoculated with *Agrobacterium rhizogenes* (strain 15834; Ri plasmid). The hairy roots produced buds on a hormone-free medium in the presence of red light, but not far-red light. Thus, phytochrome appeared to be involved in this phenomenon. Excised roots from nontransformed plants did not exhibit this response (Saitou, Kamada, & Harada, 1990).

For additional information on organogenesis in cultures initiated from certain major plant families (Solanaceae, Cruciferae, Leguminosae, Compositae, and Graminaceae), the review by Flick et al. (1983) will be a helpful starting point in devising appropriate growth regulator concentrations.

In the present experiment the student will attempt to induce the formation of plantlets from explants of *Saintpaulia ionantha* Wendl. (African violet; Gesneriaceae). This plant has been propagated in vitro from explants of leaf lamina (Start & Cummings, 1976; Cooke, 1977; Vasquez, Davey, & Short, 1977), petioles (Bilkey, McCown, & Hildebrandt, 1978; Harney & Knap, 1979), and floral organs (Hughes, 1977; Vasquez et al., 1977). Harney (1982) has given a review of the in vitro propagation procedures for the regeneration of African violets.

Obtain from the florist a healthy African violet plant with an abundance of dark-green foliage. Explants will be prepared from the youngest leaves by transversely slicing the petioles into segments approximately 10 mm in length. In addition, prepare some explants from the base of the lamina, that is, at the point of attachment of the petiole to the blade of the leaf. Because of the fuzzy texture of the epidermal layer, a wetting agent must be used during the sterilization procedure. The objectives of the experiment are twofold:

1. three different concentrations of benzylaminopurine will be tested for shoot initiation (cytokinin:auxin ratios of 5:1, 10:1, and 50:1), and
2. regeneration of African violet plants will be accomplished by rooting the newly formed shoots.

### LIST OF MATERIALS

*Sterilization mode:* C, chemical; O, oven; A, autoclave

- C    a mature, healthy African violet plant; each student will need 10 leaves
- –    200 cm$^3$ aqueous solution (20% v/v) bleach containing a final concentration of approximately 1.0% (v/v) NaOCl; add a few drops of Tween 20
- O    250 cm$^3$ beakers (four)
- O    stainless steel forceps; place in test tube and wrap with aluminum foil prior to sterilization
- O    culture tubes, 21 × 70 mm (15); cap with foil, place in Pyrex storage jars, and wrap entire unit with foil
- A    9-cm Petri dishes, each containing two sheets Whatman No. 1 filter paper (10); enclose in paper bag
- A    125-cm$^3$ Erlenmeyer flasks, each containing 100 cm DDH$_2$O (six)
- A    150 cm$^3$ MS medium; the experiment involves three different benzylaminopurine concentrations, so prepare 3 × 50 cm$^3$ aliquots in 125-cm$^3$ Erlenmeyer flasks
- C    scalpel; Bard–Parker No. 7 surgical knife handle with a No. 10 blade is recommended
- –    ethanol (80% v/v) in a metal can
- –    ethanol (70% v/v) in plastic squeeze bottle
- –    methanol lamp
- –    interval timer
- –    heavy-duty aluminum
- –    plant growth chamber (25–29 °C) with illumination
- –    pots; sterile soil mixture

# Organogenesis

## PROCEDURE

Prepare 1 liter of MS medium as outlined in Chapter 4, and supplement with *myo*-inositol (100 mg/l), nicotinic acid (0.5 mg/l), pyridoxine-HCl (0.5 mg/l), and thiamine-HCl (0.4 mg/l). One liter will be sufficient for six students. Prepare three 50-cm$^3$ aliquots. Each aliquot will contain sucrose (2.0% w/v), Gelrite (0.2% w/v), $\alpha$-NAA (0.1 mg/l), and a supplement of benzylaminopurine (BA). In aliquot 1 add 0.5 mg/l BA; in aliquot 2 add 1.0 mg/l BA; and in aliquot 3 add 5.0 mg/l BA. Using a magnetic stirrer adjust the pH of each aliquot to 5.7. The autoclaved medium is dispensed into 15 culture tubes (five tubes per treatment; 10 cm medium in each tube).

*A brief lesson in stock preparation.* The three aliquots will require the following quantities of BA:

1. 0.5 mg/l = 0.025 mg/50 ml;
2. 1.0 mg/l = 0.05 mg/50 ml; and
3. 5.0 mg/l = 0.25 mg/50 ml.

Because of the wide range of concentrations, it is advisable to prepare two stock solutions. Prepare stock A with 25 mg BA/100 ml DDH$_2$O in a volumetric flask. One milliliter will contain 0.25 mg and should be added to aliquot 3. Prepare stock B by diluting stock A by one-tenth, that is, 10 ml stock A to 90 ml DDH$_2$O. Stock B now has a concentration of 0.025 mg BA/ml. Add 1 ml stock B to aliquot 1, and 2 ml of stock B to aliquot 2. (Review Chapter 4 on media preparation.)

### Culture procedure

1. Excise 10 small-to-medium-sized leaves, including the petioles, and wash the leaves briefly in cool, soapy water. Rinse in running tap water, and prepare for aseptic procedures. Because the steps in the preparation of the plant material for organogenesis follow the same basic procedure as outlined in Chapter 5, the following instructions will be given in abbreviated form.

2. Immerse the leaves in a 20% (v/v) aqueous solution of Clorox or other commercial bleach for 10 min. The bleach solution should contain a few drops of Tween 20. Rinse the leaves in three

successive baths of $DDH_2O$ (200 $cm^3$ each). Each rinse should last about 30 sec to 1 min.

3. Each leaf is transferred to a sterile Petri dish containing filter paper, and explants are prepared with the aid of forceps and scalpel. The filter paper will remove the excess moisture from the rinse water. Petiole segments about 1 cm in length make excellent explants. Also prepare 1-$cm^2$ explants from the lamina tissues located near the point of attachment of the petiole and from the center of the blade. Position the petiole explants flat, that is, parallel to the surface of the medium. The laminar explants should be positioned with the lower epidermis touching the surface of the medium.

4. Place the cultures in a growth chamber maintained at 25–29 °C with 16-hr photoperiods furnished by a combination of Gro-Lux and cool-white fluorescent tubes. The light intensity should be about 1,000–1,500 lux.

## RESULTS

Shoots will appear within about four weeks, and after about eight weeks of culture the regenerated shoots will be ready to be aseptically subdivided and subcultured for the initiation and development of root systems. Rooting is promoted by transferring the shoots to a fresh medium that is devoid of plant growth regulators and has a sucrose concentration of about 1.6% (w/v) (Start & Cummings, 1976). This subculture step may be unnecessary since the subdivided shoots apparently can establish a root system in a sterile potting soil mixture (Harney & Knap, 1979). A small bag of sterile African violet soil mixture can be purchased from a local supermarket or florist shop. The miniature pots should be maintained under a relatively high humidity with adequate lighting. Direct sunlight, however, can be harmful. The requirements for hardening off the plantlets are similar to those for plantlets regenerated by the micropropagation of the shoot apex (Chapter 10).

## QUESTIONS FOR DISCUSSION

1. What are the main tissue culture methods of plant regeneration?
2. Is an embryo a plant organ? Explain.

3. What was Skoog's hypothesis in regard to organogenesis? Is it used today? Can you think of some reasons why the application of Skoog's hypothesis may fail to give the expected results?
4. Why does a petiole explant from an immature African violet leaf give a better organogenetic response than an explant taken from the tip of a mature leaf?

APPENDIX

*Shoot initiation on stem explants of Populus* (R. R. Willing, pers. commun.). Remove a first-year twig from a poplar (*Populus* sp.) tree. Prepare an internodal explant 1–2 cm in length and split the stem segment lengthwise. Place the segments, arranged with the cut surfaces of the bark uppermost, in a Petri dish containing two sheets of Whatman No. 1 filter paper wetted with an aqueous solution of kinetin (2.0 mg/l). Seal the plate with Parafilm M, or rewet the filter paper periodically with the kinetin solution. Incubate at room temperature in the laboratory. After a few weeks callus will be evident, and shortly thereafter bud formation will occur. This is a simple experiment to perform because aseptic conditions are not necessary. The student should attempt to induce root formation and the regeneration of poplar plantlets. The production of callus by isolated segments of poplar stem "cultured" on wet filter paper was first observed by Rechinger in 1893.

SELECTED REFERENCES

Aartrijk, J. Van, & Blom-Barnhoorn, G. J. (1983). Adventitious bud formation from bulb-scale explants of *Lilium speciosum* Thunb. *In vitro*. Effect of wounding, TIBA, and temperature. *Z. Pflanzenphysiol.* 110, 355–63.

Bilkey, P. C., McCown, B. H., & Hildebrandt, A. C. (1978). Micropropagation of African violet from petiole cross sections. *HortScience* 13, 37–8.

Chlyah, A., & Tran Thanh Van, M. (1975). Differential reactivity in epidermal cells of *Begonia rex* excised and grown in vitro. *Physiol. Plant.* 35, 16–20.

Cooke, R. C. (1977). Tissue culture propagation of African violets. *HortScience* 12, 549.

Dixon, R. A. (1985). Isolation and maintenance of callus and cell suspension cultures. In *Plant cell culture: A practical approach*, ed. R. A. Dixon, pp. 1–20. Oxford: IRL Press.

Everett, N. (1982). The determination phase of differentiation. In *Plant tissue culture 1982*, ed. A. Fujiwara, pp. 93–4. Tokyo: Japanese Association for Plant Tissue Culture.

Flick, C. E., Evans, D. A., & Sharp, W. R. (1983). Organogenesis. In *Handbook of plant cell culture*, vol. 1, *Techniques for propagation and breeding*, ed. D. A. Evans, W. R. Sharp, P. V. Ammirato, & Y. Yamada, pp. 13–81. New York: Macmillan.

Gresshoff, P. M. (1978). Phytohormones and growth and differentiation of cells and tissues cultured in vitro. In *Phytohormones and related compounds – A comprehensive treatise*, ed. D. S. Letham, P. B. Goodwin, & T. J. V. Higgins, vol. 2, pp. 1–29. Amsterdam: North-Holland.

Harney, P. M. (1982). Tissue culture propagation of some herbaceous horticultural plants. In *Application of plant cell and tissue culture, agriculture and industry*, ed. D. T. Tomes, B. E. Ellis, P. M. Harney, K. J. Kasha, & R. L. Peterson, pp. 187–208. Guelph: University of Guelph Press.

Harney, P. M., & Knap, A. (1979). A technique for the *in vitro* propagation of African violets using petioles. *Can. J. Plant Sci.* 59, 263–6.

Hughes, K. W. (1977). Tissue culture of haploid and diploid African violets (*Saintpaulia ionantha*). *In Vitro* 13, 169.

Huxter, T. J., Reid, D. M., & Thorpe, T. A. (1979). Ethylene production by tobacco (*Nicotiana tabacum*) callus. *Physiol. Plant.* 46, 374–80.

Rechinger, C. (1893). Untersuchungen über die Grenzen der Teilbarkeit im Pflanzenreich. *Abh. Zool.-Bot. Ges. (Vienna)* 43, 310–34.

Saitou, T., Kamada, H., & Harada, H. (1990). Light requirement for adventitious bud formation in hairy roots of horseradish. *Abstracts VIIth International Congress Plant Tissue and Cell Culture* (abst. B6–76). Amsterdam: IAPTC.

Skoog, F. (1944). Growth and organ formation in tobacco tissue cultures. *Am. J. Bot.* 31, 19–24.

Skoog, F., & Miller, C. O. (1957). Chemical regulation of growth and organ formation in plant tissues cultured in vitro. *Symp. Soc. Exp. Biol.* 11, 118–30.

Skoog, F., & Tsui, C. (1948). Chemical control of growth and bud formation in tobacco stem segments and callus cultured in vitro. *Am. J. Bot.* 35, 782–7.

Start, N. D., & Cummings, B. G. (1976). *In vitro* propagation of *Saintpaulia ionantha* Wendl. *HortScience* 11, 204–6.

Street, H. E. (1977). The anatomy and physiology of morphogenesis: Studies involving tissue and cell cultures. In *La culture des tissus et des cellules des végétaux: Résultats généraux et réalisations practiques*, ed. R. J. Gautheret, pp. 20–33. Paris: Masson.

Thorpe, T. A. (1980). Organogenesis in vitro: Structural, physiological, and biochemical aspects. In *Perspectives in plant cell and tissue culture. Int. Rev. Cytol. Suppl. 11A*, ed. I. K. Vasil, pp. 71–111. New York: Academic Press.

Tran Thanh Van, K. (1980a). Control of morphogenesis by inherent and exogenously applied factors in thin cell layers. In *Perspectives in plant cell and tissue culture. Int. Rev. Cytol. Suppl. 11A*, ed. I. K. Vasil, pp. 175–94. New York: Academic Press.

(1980b). Control of morphogenesis of what shapes a group of cells? In *Advances in Biochemical Engineering*, ed. A. Fiechter, vol. 18, pp. 151–71. Berlin: Springer–Verlag.

Tran Thanh Van, M., Chlyah, H., & Chlyah, A. (1974). Regulation of organogenesis in thin layers of epidermal and sub-epidermal cells. *Tissue culture and plant science 1974*, ed. H. E. Street, pp. 101–39. London: Academic Press.

Tran Thanh Van, K., & Mutaftschiev, S. (1990). Signals influencing cell elongation, cell enlargement, cell division and morphogenesis. In *Progress in plant cellular and molecular biology*, ed. H. J. J. Nijkamp, L. H. W. van der Plas, & J. Van Aartrijk, pp. 514–19. Dordrecht: Kluwer Academic Publishers.

Tran Thanh Van, K., & Trinh, H. (1978). Morphogenesis in thin cell layers: Concept, methodology and results. In *Frontiers of plant tissue culture 1978*, ed. T. A. Thorpe, pp. 37–48. Calgary: IAPTC.

Vasquez, A. M., Davey, M. R., & Short, K. C. (1977). Organogenesis in cultures of *Saintpaulia ionantha*. *Acta Hortic.* 78, 249–58.

Welander, M., & Huntrieser, I. (1981). The rooting ability of shoots raised "in vitro" from the apple rootstock A2 in juvenile and in adult growth phase. *Physiol. Plant.* 53, 301–6.

# 7

## Cell suspensions

The culture of cell suspensions is as close as cell culture comes to the fermentation biology of microbial systems.

According to King (1980) the term "suspension culture" has no clear-cut biological definition, and such tissue culture systems are evidently more than simply aggregates of cells suspended in a liquid medium. A suspension culture originates with a "random critical event" occurring during the early exposure of the plant cells to the liquid medium. Cells undergoing this transition in metabolism and growth rate produce a "cell line." Some of the characteristics of cell lines include the following (King, 1980):

1. a high degree of cell separation,
2. homogeneous cell morphology,
3. distinct nuclei and dense cytoplasm,
4. starch granules,
5. relatively few tracheary elements,
6. doubling times of 24–72 hr,
7. loss of totipotency,
8. hormone habituation, and
9. increased ploidy levels.

Suspension cultures vary considerably in the expression of these and other cell line characteristics, and consequently these systems remain an ill-defined group.

Cell suspension cultures are generally initiated by transferring fragments of undifferentiated callus to a liquid medium, which is then agitated during the culture period. Although a longer time is required, suspension cultures can be started by inoculating the liquid medium with an explant of differentiated plant material (e.g., a fragment of hypocotyl or cotyledon). The dividing cells will gradually free themselves from the inoculum because of the swirling action of the liquid. It should be kept in mind, however, that no suspension culture has been shown to be composed entirely of single cells (Butcher & Ingram, 1976). After a short time the culture will be composed of single cells, cellular aggregates of various sizes, residual pieces of the inoculum, and the remains of dead cells. The term "friability" is used to describe the separation of cells following cell division. Formation of a "good suspension" (i.e., a culture consisting of a high percentage of single cells and small clusters of cells) is much more complex than finding the optimum environmental conditions for cell separation (King, 1980). The degree of cell separation of established cultures already having the characteristics of high friability can be modified by changing the composition of the nutrient medium. Increasing the auxin:cytokinin ratio will, in some cases, produce a more friable culture. On the other hand, some cultures exhibit low friability regardless of cultural conditions (King & Street, 1977). There is no standard procedure that can be recommended for starting cell suspension cultures from callus; the choice of suitable conditions is largely determined by trial and error (King & Street, 1977).

The initiation of a cell suspension culture requires a relatively large amount of callus to serve as the inoculum, for example, approximately 2–3 g for 100 $cm^3$ (Helgeson, 1979). When the plant material is first placed in the medium there is an initial lag period prior to any sign of cell division (Fig. 7.1). This is followed by an exponential rise in cell number and a linear increase in the cell population. There is a gradual deceleration in the division rate. Finally, the cells enter a stationary or nondividing stage. In order to maintain the viability of the culture, the cells should be subcultured early during this stationary phase.

Because cells from different plant material vary in the length of time they remain viable during the stationary phase, it may be prudent to subculture during the period of progressive deceleration. Passage time can be learned only from experience, and a given suspension culture should be subcultured at a time approximating the maximum cell density. For many suspension cultures the maximum cell density is reached within about 18–25 days, although the passage time for some extremely active cultures may be as short as 6–9 days (Street, 1977). At the time of the first subculture it will be necessary to filter the culture through a nylon net or stainless steel filter to remove the larger cell aggregates and residual inoculum that would clog the orifice of a pipette. A small sample should be withdrawn and the cell density determined before subculturing. There is a critical cell density below which the culture will not grow; for example, this value is 9–15 × $10^3$ cells/$cm^3$ for a clone of sycamore cells (*Acer pseudoplatanus*) (Street, 1977).

Cell suspension cultures must be agitated or subjected to forced aeration, and a platform (orbital) shaker is used for this purpose in most laboratories. The best speed range for cultures in 250-$cm^3$ Erlenmeyer flasks is 100–120 rpm (Thomas & Davey, 1975). The volume of liquid in relation to the size of the flask is important for adequate aeration (i.e., the liquid medium should occupy about 20% of the total volume of the flask). Other devices for aeration include magnetic stirrers, roller cultures, and Steward's auxophy-

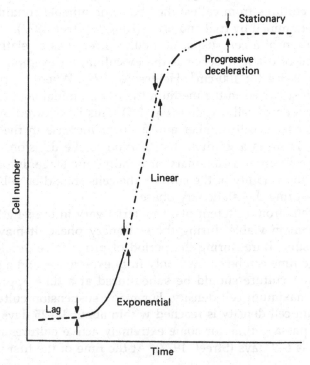

Figure 7.1. Growth curve of a cell suspension grown under batch conditions relating total cell number per unit volume to time.

ton. The latter apparatus slowly rotates the cultures in nipple flasks and tumble tubes. Microcultures do not need any device for oxygenation of the nurtured cells.

There is some terminology associated with cell suspension cultures. The present experiment involves the preparation of a "batch culture," defined as a culture grown in a fixed volume of culture medium. Our experiment is also a "closed culture" because all cells are retained, and a continual increase in cell density will occur until the stationary phase is reached. A "closed continuous system" involves a continuous influx of fresh medium and a withdrawal of spent medium. An "open continuous system" is similar to the closed in the replenishment of the nutrient medium; in addition, however, the cells are harvested. Examples of open continuous systems are "chemostats" and "turbidostats." In a chemostat the continuous flow of fresh medium into the system is set at a predetermined rate; this influx of nutrients will largely determine the growth rate of the culture. In the turbidostat, cell density is set at some predetermined level, and fresh medium is added periodically to maintain that density within the preset limits. Cell density in the turbidostat is determined with a photocell control device.

The systems for culture of cell suspensions have recently expanded to pilot plant and semicommercial scale.

In the present experiment it is suggested that the callus culture described in Chapter 5 be used as the inoculum for the cell suspension culture. Carrot callus is an excellent inoculum for a suspension culture.

A detailed account of the nutritional requirements of cell suspension cultures is beyond the scope of this introduction; however, some specific refinements in the nutritional needs of cultures are discussed in the review by Ojima and Ohira (1978). The process can be scaled up to industrial level using equipment similar to that in Figure 7.2.

### LIST OF MATERIALS

*Sterilization mode:* C, chemical; O, oven; A, autoclave

- – actively growing callus cultures, preferably derived from carrot taproot (*Daucus carota* L.); 5–10 culture tubes
- O forceps

C scalpel
A 9-cm Petri dishes, each containing two sheets Whatman No. 1 filter paper (two)
A 125-cm$^3$ Erlenmeyer flasks, each containing 25 cm$^3$ liquid MS medium, 2,4-D (1 µg/cm$^3$), and sucrose (750 mg) (five)
A 100-µm pore size nylon mesh filtration cloth equipped with a Hirsh funnel and flask
A syringe equipped with wide-bore cannula, preferably designed to deliver a preset volume

Figure 7.2. An industrial pilot plant fermenter for plant cells. (Photo courtesy of M. J. C. Rhodes and P. D. G. Wilson, Institute of Food Research, Norwich Laboratory.)

A  Pasteur pipette or 1 cm$^3$ pipette for sampling culture
- ethanol (80% v/v) dip
- ethanol (70% v/v) in plastic squeeze bottle
- aqueous solution chromium trioxide (8% w/v)
- hot plate
- Pasteur pipette for sample maceration
- Sedgwick–Rafter slide or hemocytometer
- depression slide; microscope slide; cover slips
- light microscope (100× magnification)
- dissecting microscope
- 125-cm$^3$ beaker
- ocular micrometer, calibrated (100×) for the light microscope (optional)

### PROCEDURE

*Carrot cell suspension culture.* Carrot cell suspension cultures can be initiated on a basal MS medium supplemented with 2,4-D (1.0 mg/l) and sucrose (Nishi & Sugano, 1970). A suspension culture of carrot does not require an exogenous cytokinin (Nishi and Sugano, 1970), and the presence of auxin as the sole plant growth regulator in the medium may improve the friability of the culture. For the present experiment each 125-cm$^3$ Erlenmeyer flask will contain 25 cm$^3$ of basal MS medium, plus 2,4-D (1$\mu$g/cm$^3$) and sucrose (750 mg).

The callus resulting from the experiment conducted in Chapter 5 should be removed from the culture tubes with forceps and transferred to a Petri dish containing Whatman No. 1 filter paper. Trim the callus blocks with the scalpel and use only the young, actively growing callus for the inoculum. Each flask should receive an inoculum of about 500–750 mg of callus in order to ensure the initiation of the culture. Brownish callus may be indicative of senescence and should be discarded. Place the inoculated flasks on the shaker and set the speed at 100 rpm. The shaker with the flasks should be placed in an air-conditioned enclosure maintained at 25–27 °C. If, during the first few days, the medium appears "milky," this is a sign that contamination occurred during the inoculation. The initial subculture can be performed after 7–10 days, although it is first necessary to filter the culture through an industrial nylon mesh filter in order to remove the residual inoculum and larger clumps of cells. This is somewhat difficult because the cell sizes in carrot suspension cultures range from 50 to 300 μm in diameter (Steward, Mapes, & Smith, 1958). A nylon mesh filter (100-μm pore size), sterilized with a Hirsh funnel, will be satisfactory. The next step is to determine the cell density of the culture. It is impossible to subculture and maintain the culture unless the cell density is within a given range. According to Street (1977), most suspension cultures contain $0.5$–$2.5 \times 10^5$ cells/cm$^3$ after dilution with the fresh medium. A sample is taken with a syringe equipped with a wide-bore cannula. Cell counting is best achieved by a closely regulated cell separation with chromium trioxide. Because the acid treatment is highly destructive, it should be long enough to achieve a reasonable degree of cell separation without destroying the sample (Street, 1977). Chromium trioxide is *highly corrosive*. Be particularly careful not to spill the acid on your skin or clothing, or to breathe the fumes. Add one volume of the cell suspension to two volumes of chromium trioxide (8% w/v), heat the mixture to 70 °C for 2–15 min inside a hood, cool it, and then macerate the sample further by pumping it repeatedly through the orifice of a Pasteur pipette. The macerate is placed in a Sedgwick–Rafter slide or hemocytometer for cell counting. Discard the acid solution with great care and rinse the Pasteur pipette and slide after use. Additional details on the quantitation of results are given in Chapter 17. The necessary volume of inoculum may be calculated in order to give a final concentration of cells within the minimum density level (i.e., about $0.5$–$2.5 \times 10^5$ cells/cm$^3$).

## RESULTS

Remove a 1-cm$^3$ sample of the cell suspension culture with a sterile Pasteur pipette and discharge the contents into a small beaker. Place about 0.1 cm$^3$ of the suspension in a depression slide and examine the preparation with a dissection microscope. Then examine a droplet of the suspension with the light microscope (100×). In what obvious ways do these cultured cells differ from the cells of the primary explant excised from the cambial zone of the carrot root? With the aid of an ocular micrometer, the approximate range of sizes of the cells can be estimated. Some of the cellular details may be enhanced by employing a biological stain to increase the contrast (Chapter 5).

One of the cell suspension cultures may be sampled daily and the cell density determined. Prepare a plot of the cell number versus time. Does the resulting curve of the growth rate approximate the curve shown in Figure 7.1?

### QUESTIONS FOR DISCUSSION

1. What are some advantages of using an aqueous medium in comparison with an agar-solidified medium? Can you think of any advantages that were not mentioned in this chapter?
2. In addition to providing a source of oxygen, what are some other possible effects of agitating a cultured plant cell or organ?
3. From the time of inoculation, what growth stages are exhibited by a cell suspension culture? Why does the rate of growth resemble an S-shaped curve (i.e., what are the reasons for each of these fluctuations in the growth curve)?
4. Define each of the following terms: batch culture, open culture.

### APPENDIX

*Viability of cell suspension.* A method has been reported for the determination of cell viability in suspension cultures (Towill & Mazur, 1975 ). Viability, in this instance, refers to the capability of the cells to exhibit cell division. The tetrazolium reagents accept electrons from the electron transport chain of the mitochondria; as a result, these oxidation–reduction indicators are converted to brightly colored formazan precipitates. There is a close, positive correlation between the amount of formazan produced and the percentage of viable cells in the sample. The technique involves the preparation of an aqueous solution of 2,3,5-triphenyltetrazolium chloride (0.8% w/v) dissolved in a mixture of buffer and suspension culture in the ratio of two parts buffer to one part suspension culture. A sodium phosphate buffer (0.05 $M$) giving a final pH of 7.5 is recommended. After an 18–20-hr incubation in the dark at room temperature, a red precipitate will be observed in the viable cells. The cells are pelleted by centrifugation and washed once with distilled water, and the dye is extracted with 3 cm$^3$ ethanol (95% v/v) for 30 min. If cell clumps are present, gentle heating will facilitate the extraction (60 °C; 5–15 min). The absorbency is determined in a spectrophotometer (485 nm).

*Experiment:* Take a sample of the carrot cell suspension culture that has been subjected to cell counting by Sedgwick–Rafter slide or hemocytometer, and determine the absorbance of the formazan dye produced by the preceding technique. Prepare serial dilutions of the sample and redetermine the formazan content of each of the dilutions. Do you find a linear relationship between cell numbers and dye production in the series of dilutions?

SELECTED REFERENCES

Butcher, D. N., & Ingram, D. S. (1976). *Plant tissue culture.* London: Arnold.
Helgeson, J. P. (1979). Tissue and cell suspension culture. In *Nicotiana: Procedures for experimental use,* ed. R. D. Durbin, pp. 52–9. Washington, D.C.: USDA Tech. Bull. No. 1586.
King, P. J. (1980). Cell proliferation and growth in suspension cultures. *Int. Rev. Cytol. Suppl. 11A,* ed. I. K. Vasil, pp. 25–54.
King, P. J., & Street, H. E. (1977). Growth patterns in cell cultures. In *Plant tissue and cell culture,* ed. H. E. Street, pp. 307–87. Oxford: Blackwell Scientific Publications.
Nishi, A., & Sugano, N. (1970). Growth and division of carrot cells in suspension culture. *Plant Cell Physiol.* 11, 757–65.
Ojima, K., & Ohira, K. (1978). Nutritional requirements of callus and cell suspension cultures. In *Frontiers of plant tissue culture 1978,* ed. T. A. Thorpe, pp. 265–75. Calgary: IAPTC.
Steward, F. C., Mapes, M. O., & Smith, J. (1958). Growth and organized development of cultured cells: I. Growth and division of freely suspended cells. *Am. J. Bot.* 45, 693–703.
Street, H. E. (1977). Cell (suspension) cultures – techniques. In *Plant tissue and cell culture,* 2d Ed., ed. H. E. Street, pp. 61–102. Oxford: Blackwell Scientific Publications.
Thomas, E., & Davey, M. R. (1975). *From single cells to plants.* London: Wykeham.
Towill, L. E., & Mazur, P. (1975). Studies on the reduction of 2,3,5-triphenyltetrazolium chloride as a viability assay for plant tissue cultures. *Can. J. Bot.* 53, 1097–102.

# 8

## Somatic embryogenesis

The capacity of flowering plants to produce embryos is not restricted to the development of the fertilized egg; embryos ("embryoids") can be induced to form in cultured plant tissues. This phenomenon was first observed in suspension cultures of carrot *(Daucus carota)* by Steward, Mapes, and Mears (1958) and in carrot callus grown on an agar medium by Reinert (1959). This is a general phenomenon in higher plants, and experimental somatic embryogenesis has been reported in tissues cultured from more than 30 plant families (Raghavan, 1976; Narayanaswamy, 1977; Ammirato, 1983; see Table 8.1).

Somatic embryoids may arise in vitro from three sources of cultured diploid cells (Kohlenbach, 1978):

1. vegetative cells of mature plants,
2. reproductive tissues other than the zygote, and
3. hypocotyls and cotyledons of embryos and young plantlets without any intervening callus development.

Precisely how these adventive embryoids arise from these tissues has been the subject of numerous studies. According to Sharp and his colleagues (1980), somatic embryogenesis may be initiated in two different ways. In some cultures embryogenesis occurs directly in the absence of any callus production from "preembryonic determined cells" that are programmed for embryonic differentiation. The second type of development requires some prior callus proliferation, and embryos originate from "induced embryogenic cells" within the callus (Sharp et al., 1980). Carrot cells are an example of the latter case. Although individual carrot cells are totipotent and carry all the genetic templates necessary for the development of the whole plant, isolated single cells do not generally become transformed into embryos by repeated divisions (McWilliam, Smith, & Street, 1974). Embryoids are initiated in callus from superficial clumps of cells associated with highly vacuolated cells that do not take part in embryogenesis. Observations have been made on the ultrastructure of embryogenic clumps of cells in callus derived from carrot (McWilliam et al., 1974; Street & Withers, 1974) and *Ranunculus* (Thomas, Konar, & Street, 1972). The embryoid-forming cells are characterized by dense cytoplasmic contents, large starch grains, and a relatively large nucleus with a darkly stained nucleolus. Staining reagents indicated that these embryogenic cells have high concentrations of protein and RNA. These cells also exhibited high dehydrogenase activity with tetrazolium staining (McWilliam et al., 1974; Street & Withers, 1974). Each developing embryoid passes through the sequential stages of embryo formation (i.e., globular, heart shape, and torpedo shape) (Fig. 8.1). Two critical events are involved in the early programming of this process (Kohlenbach, 1978):

Table 8.1. *Examples of plants in which somatic embryogenesis has been induced under in vitro conditions*

| Plant | Reference |
|---|---|
| *Bromus inermis* Leyss. var. Machar | Gamborg, Constabel, & Miller, 1970 |
| | Constabel, Miller, & Gamborg, 1971 |
| *Brassica oleracea* var. Borytis | Pareek & Chandra, 1978 |
| *Atropa belladonna* L. | Konar, Thomas, & Street, 1972 |
| *Carica papaya* | Litz & Conover, 1983 |
| *Manihot esculentum* | Stamp & Henshaw, 1982 |
| *Coffea arabica* L. | Sondahl & Sharp, 1977 |
| *Pinus ponderosa* Dougl. | Moore, 1976 |
| *Solanum melongena* L. | Yamada, Nakagawa, & Sinoto, 1967 |
| *Daucus carota* L. | Steward et al., 1958 |
| | Reinert, 1959 |
| | Halperin & Wetherell, 1964 |
| | Homès, 1967 |
| *Citrus sinesis* var. "Shamouti" orange | Kochba & Button, 1974 |
| *Hordeum vulgare* | Bayliss & Dunn, 1979 |
| *Ranunculus sceleratus* L. | Nataraja & Konar, 1970 |
| *Saccharum officinarum* | Jane Ho & Vasil, 1983 |
| *Nicotiana tabacum* var. Samsun | Lörz, Potrykus, & Thomas, 1977 |

1. the induction of cytodifferentiation of the proembryoid cells, and
2. the unfolding of the developmental sequence by these proembryoid cells.

Although a given culture may differentiate these embryogenic cells, their further development may be blocked by an imbalance of chemicals in the culture medium. Abnormalities, known as "embryonal budding" and "embryogenic clump formation," may occur if relatively high levels of auxin are present in the medium after the embryogenic cells have been differentiated (Kohlenbach, 1978). In other words, two distinctly different types of media may

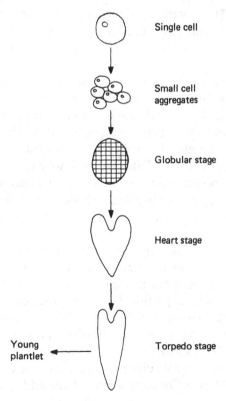

Figure 8.1. Stages of somatic embryogenesis. Following repeated cell divisions, cell aggregates progressively develop and pass through globular, heart, and torpedo stages before ultimately forming plantlets.

be required: one medium for the initiation of the embryonic cells and another for the subsequent development of these cells into embryoids. The first (induction) medium must contain auxin. The second generally consists of a mixture either lacking auxin, with a lower concentration of the same auxin, or with reduced levels of a different auxin. With some plants, however, both embryo initiation and subsequent maturation occur on the first medium, and a second medium is employed for plantlet development (Ammirato, 1983).

The most important chemical factors involved in the induction medium are auxin and reduced nitrogen. Substantial amounts of reduced nitrogen are required in both the first and second media (Ammirato, 1983). In wild carrot cultures the addition of 10 m$M$ $NH_4Cl$ to an embryogenic medium already containing $KNO_3$ (12–40 m$M$) produced near-optimal numbers of embryoids. Glutamine, glutamic acid, urea, and alanine, respectively, were found partially to replace $NH_4Cl$ as a supplement to $KNO_3$ (Wetherell & Dougall, 1976). These various nitrogen sources are not specific for the induction of embryogenesis, although at low concentrations organic forms are much more effective than inorganic nitrogen compounds.

The role of cytokinins in embryogenesis is somewhat obscure because of conflicting results. Although zeatin (0.1 $\mu M$) stimulates embryogenesis in carrot cell suspensions during the auxin-free subculture (secondary culture), the process is inhibited by the addition of either kinetin or benzylaminopurine to the medium (Fujimura & Komamine, 1975). The inhibitory effect of exogenous cytokinins may result from the increase in endogenous cytokinins in the developing embryoids (Al-Abta & Collin, 1979). Further information on cytokinins can be found in the review by Ammirato (1983).

Supplementing the medium with activated charcoal has facilitated embryogenesis in several cultures. The induction of embryogenesis was successful in *Daucus carota* cultures containing charcoal when auxin depletion failed to produce the desired results (Fridborg & Eriksson, 1975; Drew, 1979). Charcoal was a requirement for embryogenesis in English ivy *(Hedera helix)* cultures (Banks, 1979). Evidence indicates that charcoal may adsorb a wide variety of inhibitory substances as well as growth promoters (Ammirato, 1983). See Chapters 4 and 11 for additional comments on charcoal.

In general, embryogenesis occurs most readily in short-term cultures, and this ability decreases with increasing duration of culture (Reinert, Bajaj, & Zbell, 1977). There are exceptions, however, and embryogenesis has been reported in some cultures maintained over a period of years. Embryoid formation begins in carrot cultures about 4 to 6 weeks after isolation of the tissues, and an optimum embryogenic potential is reached after about 15 weeks (Reinert, Backs-Hüsemann, & Zerman, 1971). After the embryogenic potential has apparently been lost following 36 weeks in vitro, the carrot cultures can once again be induced to produce embryoids by transfer to an appropriate medium (Reinert et al., 1971). This temporary loss of embryogenic potential presumably results from the lack of biosynthesis of certain "embryogenic substances" by the cultured cells. In addition, changes in ploidy of the cultured cells may lead to a loss of morphogenetic potential (Smith & Street, 1974).

A technique has been developed for the physical separation of the globular, heart, and torpedo stages of embryogenesis by using glass beads to screen the cultures (Warren & Fowler, 1977). This procedure should prove useful for further biochemical studies of the developmental process.

Some progress has been made in inducing synchronization of somatic embryogenesis. A high degree of synchronization of embryogenesis was achieved in a carrot suspension culture by:

1. sieving the initial cell populations,
2. employing density gradient centrifugation in Ficoll solutions, and
3. using repeated low-speed centrifugation for 5-sec periods.

The resulting cell clusters, cultured in an auxin-free medium containing zeatin, gave a greater than 90% frequency of embryoid formation (Fujimura & Komamine, 1979).

Although the regeneration of whole plants by embryogenesis has been relatively rare in the Gramineae, somatic embyros have been formed directly from leaf mesophyll cells of orchard grass (*Dactylis glomerata* L.) without an intervening callus. Explants, prepared from the basal portions of two innermost leaves, were cultured on a Schenk and Hildebrandt medium containing 30 $\mu M$ 3,6-dichloro-O-anisic acid (dicamba) and 0.8% (w/v) agar. Plantlet formation occurred after subculturing the embryos on the same medium lacking dicamba (Conger et al., 1983).

Endogenous polyamines appear to be required for the induction of embryogenesis in cultures of wild carrot (Feirer, Mignon, & Litvay, 1984). Embryogenic cultures of *Daucus carota* treated with a specific inhibitor of arginine decarboxylase showed a sharp reduction in embyro production compared to untreated controls. The cultures containing the inhibitor also had relatively low levels of the polyamines putrescine and spermidine. Supplementing the culture medium with either putrescine, spermidine, or spermine restored embryogenesis to the inhibitor blocked cultures (Feirer et al., 1984).

In recent years the multiplication of somatic embryos in bioreactors has begun for commercial micropropagation. The U.S. biotechnology company DNA Plant Technology (DNAP) has applied bioreactor micropropagation via embryogenesis to coffee, banana, and pineapple (Sondahl, pers. commun.).

## LIST OF MATERIALS

*Sterilization mode:* C, chemical; O, oven; A, autoclave

- − suspension culture of carrot (*Daucus carota*) cells growing in an MS medium supplemented with 2,4-D (1.0 mg/l) and sucrose (3% w/v) (see Procedure)
- A 250-$cm^3$ Erlenmeyer flask containing 200 $cm^3$ of culture medium A, consisting of MS salts, zeatin (0.2 mg/l), and 2,4-D (0.1 mg/l)
- A 250-$cm^3$ Erlenmyer flask containing 200 $cm^3$ of culture medium B, consisting of MS salts and zeatin (0.2 mg/l)
- A 125-$cm^3$ Erlenmeyer flask containing 100 $cm^3$ $DDH_2O$
- C scapel
- O forceps
- − 9-cm disposable Petri dishes, plastic, commercially sterilized (10)
- − black card with 1-cm grid lines
- A 2-$cm^3$ pipettes (six)
- − Parafilm sealing tape (one roll)
- − microscope slides; cover slips
- − light microscope (100× magnification)
- − dissecting microscope (optional)
  (Subsequent plantlet development requires filter-paper bridge apparatus and small pots containing sterile soil mixture.)

## PROCEDURE

The preparation for this experiment starts with an actively proliferating callus initiated from the taproot of carrot as described in the Appendix to Chapter 5. According to Reinert and co-workers (1971), carrot callus does not develop the potential for embryogenesis until the culture has been growing for a minimum of four to six weeks after explant isolation. Healthy fragments of the callus are transferred to a liquid culture medium (Chapter 7, Procedure), and a carrot cell suspension is initiated. This liquid medium consists of basal MS salts supplemented with 2,4-D (1.0 mg/l) and sucrose (3% w/v).

Embryoids are initiated in the present experiment in the following manner:

1. Ten Petri dishes are prepared as follows:

(a) five plates contain MS salts, zeatin (0.2 mg/l), 2,4-D (0.1 mg/l), sucrose (2% w/v), and agar (1% w/v); and
(b) five plates contain MS salts, zeatin (0.2 mg/l), sucrose (2% w/v), and agar (1% w/v).

The latter plates do not contain a source of exogenous auxin.

2. Aliquots (2 cm$^3$) of carrot suspension culture are added by pipette to the surface of the medium in the Petri dishes. The dishes are sealed with Parafilm and incubated at 25 °C in the dark for two to three weeks.

3. The test for embryogenic potential is based on a visual count of the embryoids. The "callus" from the agar surface is gently dispersed in DDH$_2$O, and the number of embryoids present is determined by placing the Petri dish over a black card marked with 1-cm$^2$ grid lines.

4. Small aliquots of this dispersed sample can be placed on a microscope slide and examined with a compound or dissecting microscope for the various stages of somatic embryogenesis.

## RESULTS

In the Petri dish containing the auxin medium the carrot cells develop into a callus and grow into small compact clumps. Embryoids, however, are not formed in these dishes. The carrot cells grown on the auxin-free medium produce large numbers of embryoids (Fig. 8.2). The embryoids are not formed in a synchronous manner: When an inoculum of this material is examined under the microscope, a wide range of developmental stages similar to those shown in Figure 8.2a–c can be seen.

After the carrot cultures have reached the late torpedo stage of development, they can be transferred to filter-paper bridges (Fig. 8.2d). (A filter-paper bridge can be made by folding a strip of filter paper [Whatman No. 1; 9 × 90 mm] in the shape of the letter "M." Its arms are immersed in the liquid medium, thus acting as a wick. The carrot emryoids are nurtured in the central "V" of the bridge.) To the culture tubes add 3 cm$^3$ of liquid medium composed of MS salts, kinetin (0.2 mg/l), and sucrose (2% w/v). The plantlets that are formed can be potted in sterile soil and grown to maturity (Fig. 8.2e).

The plantlets must be maintained under a high relative humidity to prevent excessive water loss (Chapter 10).

## QUESTIONS FOR DISCUSSION

1. What difficulties may be encountered with clonal propagation of plants by means of somatic embryogenesis?
2. What is the meaning of the term "totipotency"?
3. What is the genetic significance of totipotency?
4. What are the sequential stages of somatic embryogenesis?

## APPENDIX

The following additional experiments may be of interest to some students.

### Modified regimes for culture of carrot embryos

1. Try a range of nitrogen:auxin ratios, and attempt to determine which combination yields the highest number of embryoids. Is there any difference between the use of an organic and inorganic nitrogen source?

2. The initial carrot cell suspension culture, which is to be plated on the embryogenic agar medium, is divided into three fractions by filtration:

(a) an unfiltered fraction (mixed-cell suspension);
(b) 75–200-μm fraction (single cells and small-cell aggregates); and
(c) 75-μm fraction (single cells).

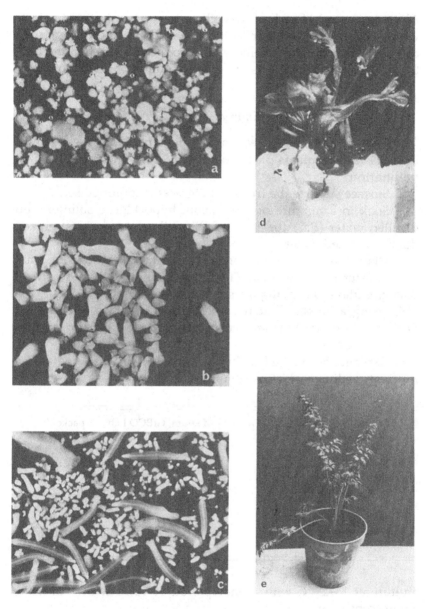

Figure 8.2. Stages of development of carrot (*Daucus carota*) embryoids: (a) young globular stage; (b) heart stage; (c) torpedo stage; (d) carrot plantlet growing on filter-paper bridge; (e) mature carrot plant derived from cultured embryoid. (Courtesy of L. A. Withers.)

Aliquots (2 cm$^3$) of each of these three fractions are plated as previously described in the procedure. After three weeks of growth, count the total number of embryoids produced by each of the fractions. Which fraction yields the greatest number of embryoids?

## Embryo rescue

1. Mature potato fruits (which look like green tomatoes) should be excised from plants over a period of 21–28 days after the first pollination.
2. Surface sterilize the fruits with 90% ethanol for 30 sec.
3. Soak in a mixture of 5% calcium hypochlorite solution and distilled water (1:1) for 15 min. Rinse three times in sterilized double-distilled water.
4. Dissect under aseptic conditions, remove seeds from fruit, and transfer to a Petri dish. (The fruit should be opened using a scalpel without damaging the seeds inside.)
5. Using a binocular stereo microscope, in an aseptic laminar flow cabinet remove the seed coats, dissect the endosperm, and excise the embryos.
6. Explant the excised embryos in each Petri dish containing nutrient medium, which is comprised as follows:

| | |
|---|---|
| Murashige Minimal (MS) Organics Medium, GIBCO Lab. | 1 packet/l |
| sucrose | 4% |
| agar | 0.8% |
| hydrolyzed casein | 1g/l |
| thiamine HC1 | 1mg/l |
| nicotinic acid | 1mg/l |
| pyridoxine | 1mg/l |
| malic acid | 100 mg/l |
| myo-inositol | 100 mg/l |
| adjusted for | pH 5.8 |

Culture at 18–22 °C with a photoperiod of 16 hr with cool-white fluorescent light.

7. After development and growth of plantlets from embryos (MSA medium), well-rooted plants are placed in small pots with moss.

(Protocol courtesy of the International Potato Center)

## SELECTED REFERENCES

Al-Abta, S., & Collin, H. A. (1979). Endogenous auxin and cytokinin changes during embryoid development in celery tissue cultures. *New Phytol. 82*, 29–35.

Ammirato, P. V. (1983). Embryogenesis. In *Handbook of plant cell culture*, vol. 1, *Techniques for propagation and breeding*, ed. D. A. Evans, W. R. Sharp, P. V. Ammirato, & Y. Yamada, pp. 82–123. New York: Macmillan.

Banks, M. S. (1979). Plant regeneration from callus of two growth phases of English ivy, *Hedera helix* L. *Z. Pflanzenphysiol. 92*, 349–53.

Bayliss, M. W., & Dunn, D. M. (1979). Factors affecting callus formation from embryos of barley *(Hordeum vulgare)*. *Plant Sci. Lett. 14*, 311–16.

Conger, B. V., Hanning, G. E., Gray, D. J., & McDaniel, J. K. (1983). Direct embryogenesis from mesophyll cells of orchardgrass. *Science 221*, 850–1.

Constabel, F., Miller, R. A., & Gamborg, O. L. (1971). Histological studies on embryos produced from cell cultures of *Bromus inermis*. *Can. J. Bot. 49*, 1415–17.

Drew, R. L. K. (1979). Effect of activated charcoal on embryogenesis and regeneration of plantlets from suspension cultures of carrot *(Daucus carota* L.). *Ann. Bot. (Lond.) 44*, 387–9.

Feirer, R. P., Mignon, G., & Litvay, J. D. (1984). Arginine decarboxylase and polyamines required for embryogenesis in the wild carrot. *Science 223*, 1433–5.

Fridborg, G., & Eriksson, T. (1975). Effects of activated charcoal on growth and morphogenesis in cell cultures. *Physiol. Plant. 34*, 306–8.

Fujimura, T., & Komamine, A. (1975). Effects of various growth regulators on the embryogenesis in a carrot cell suspension culture. *Plant Sci. Lett. 5*, 359–64.

(1979). Synchronization of somatic embryogenesis in a carrot cell suspension culture. *Plant Physiol. 64*, 162–4.

Gamborg, O. L., Constabel, F., & Miller, R. A. (1970). Embryogenesis and production of albino plants from cell cultures of *Bromus inermis*. *Planta 95*, 355–8.

Halperin, W., & Wetherell, D. F. (1964). Adventive embryony in tissue cultures of the wild carrot, *Daucus carota*. *Am. J. Bot. 51*, 274–83.

Homès, J. (1967). Induction de plantules dans des cultures in vitro de tissu de carotte. *Soc. Biol. 161*, 730–2.

Jane Ho, W., & Vasil, I. K. (1983). Somatic embryogenesis in sugarcane *(Saccharum officinarum)*. Growth and plant regeneration from embryogenic cell suspension cultures. *Ann. Bot. (Lond.) 51*, 719–26.

Kochba, J., & Button, J. (1974). The stimulation of embryogenesis and embryoid development in habituated ovular callus from the "Shamouti" orange *(Citrus sinesis)* as affected by tissue age and sucrose concentration. *Z. Pflanzenphysiol. 73*, 415–21.

Kohlenbach. H. W. (1978). Comparative somatic embryogenesis. In *Frontiers of plant tissue culture 1978*, ed. T. A. Thorpe, pp. 59–66. Calgary: IAPTC.
Konar, R. N., Thomas, E., & Street, H. E. (1972). The diversity of morphogenesis in suspension cultures of *Atropa belladonna* L. *Ann. Bot. (Lond.) 36*, 123–45.
Litz, R. E., & Conover, R. A. (1983). High frequency somatic embryogenesis from *Carica* cell suspension cultures. *Ann. Bot. (Lond.) 51*, 683–6.
Lörz, H., Potrykus, I., & Thomas, E. (1977). Somatic embryogenesis from tobacco protoplasts. *Naturwissenschaften 64*, 439–40.
McWilliam, A. A., Smith, S. M., & Street, H. E. (1974). The origin and development of embryoids in suspension cultures of carrot *(Daucus carota)*. *Ann. Bot. (Lond.) 38*, 243–50.
Moore, M. B. (1976). Early development of ponderosa pine embryos on a defined culture medium. *Silvae Genetica 25*, 23.
Narayanaswamy, S. (1977). Regeneration of plants from tissue cultures. In *Applied and fundamental aspects of plant cell, tissue, and organ culture*, ed. J. Reinert & Y. P. S. Bajaj, pp. 179–206. Berlin: Springer–Verlag.
Nataraja, K., & Konar, R. N. (1970). Induction of embryoids in reproductive and vegetative tissues of *Ranunculus sceleratus* L. in vitro. *Acta Bot. Neerl. 19*, 706–16.
Pareek, L. K., & Chandra, N. (1978). Somatic embryogenesis in leaf callus from cauliflower *Brassica oleracea* var. Borytis. *Plant Sci. Lett. 11*, 311–16.
Raghavan, V. (1976). *Experimental embryogenesis in vascular plants*. New York: Academic Press
Reinert, J. (1959). Über die Kontrolle der Morphogenese und die Induktion von Adventiveembryonen an Gewebekulturen aus Karotten. *Planta 53*, 318–33.
Reinert, J., Backs-Hüsemann, D., & Zerman, H. (1971). Determination of embryo and root formation in tissue cultures from *Daucus carota*. In *Les cultures de tissus de plantes*, pp. 261–8. Paris: Colloques Internationaux du CNRS No. 193.
Reinert, J., Bajaj, Y. P. S., & Zbell, B. (1977). Aspects of organization: Organogenesis, embryogenesis, cytodifferentiation. In *Plant tissue and cell culture*, 2d Ed., ed. H. E. Street, pp. 389–427. Oxford: Blackwell Scientific Publications.
Sharp, W. R., Sondahl, M. R., Caldas, L. S., & Maraffa, S. B. (1980). The physiology of in vitro asexual embryogenesis. *Hortic. Rev. 2*, 268–310.
Smith, S. M., & Street, H. E. (1974). The decline of embryogenic potential as callus and suspension cultures of carrot *(Daucus carota L.)* are serially subcultured. *Ann. Bot. (Lond.) 38*, 223–41.
Sondahl, M. R., & Sharp, W. R. (1977). High frequency induction of somatic embryos in cultured leaf explants of *Coffea arabica* L. *Z. Pflanzenphysiol. 81*, 395–408.

Stamp, J. A., & Henshaw, G. G. (1982). Somatic embryos in cassava. Z. Pflanzenphysiol. 105, 183–7.

Steward, F. C., Mapes, M. O., & Mears, K. (1958). Growth and organized development of cultured cells: II. Organization in cultures grown from freely suspended cells. Am. J. Bot. 45, 705–8.

Street, H. E., & Withers, L. A. (1974). The anatomy of embryogenesis in culture. In Tissue culture and plant science, 1974, ed. H. E. Street, pp. 71–100. London: Academic Press.

Thomas, E., Konar, R. N., & Street, H. E. (1972). The fine structure of embryogenic callus of Ranunculus scleratus L. J. Cell Sci. 11, 95–109.

Warren, G. S., & Fowler, M. H. (1977). A physical method for the separation of various stages in the embryogenesis of carrot cell culture. Plant Sci. Lett. 9, 71–6.

Wetherell, D. F., & Dougall, D. K. (1976). Sources of nitrogen supporting growth and embryogenesis in cultured wild carrot tissue. Physiol. Plant. 37, 97–103.

Yamada, Y., Nakagawa, H., & Sinoto, Y. (1967). Studies on the differentiation in cultured cells: I. Embryogenesis in three strains of Solanum callus. Bot. Mag. Tokyo 80, 68–74.

PART III

*Experimental: Culture of organs and organized systems*

# 9

## Isolated roots

Liquid media have been employed in the culture of excised roots and cell suspensions. In addition, various kinds of microcultures, for example, hanging-drop preparations and protoplast cultures, are special applications of liquid media. Aside from experiments on the formation and development of embryoids, relatively little has been done on the cultivation of organized plant structures other than roots in liquid media. Special attention has been given recently to the potential of hairy-root cultures for the production of secondary metabolites. A liquid medium has certain advantages over the use of nutrients in a gel-solidified matrix. When a callus is grown on a semisolid medium, diffusion gradients of nutrients and gases within the callus will lead to moderate growth and metabolism. The matrix itself may release contaminants to the culture medium, and metabolites secreted by the growing callus will accumulate in the gel matrix.

One consideration that must be given to plant material growing in a liquid medium is the availability of oxygen and the extent to which the isolated cells or organs require agitation or forced aeration. In regard to root cultures, the importance of aeration is unclear. Although Street (1957, 1969) has stated that the availability of oxygen is not a limiting factor in tomato root cultures, this view disagrees with the findings of Said and Murashige (1979). These workers found that continuous and gentle agitation of tomato root cultures resulted in a doubling of root elongation compared to root growth in stationary cultures, and the production and elongation of lateral roots were considerably improved.

There are certain advantages in achieving the continuous culture of an isolated root of a plant. This technique provides information on the nutritional requirements of the root – that is, removed from the interchange of compounds with other plant organs. We have a detailed account of the nutritional requirements for isolated tomato roots (Street, 1957). In addition to nutritional studies, root cultures from herbaceous species have provided experimental material for investigations on lateral root and bud formation, initiation of cambial activity, and nodulation (Torrey, 1965). Another advantage of using sterile cultures is the elimination of the complicating effects of microorganisms. Root clones have a rapid growth rate, and there are no difficulties in multiplying the clone in order to yield any desired quantity of plant material (Butcher & Street, 1964).

The first successful organ culture, that is, potentially unlimited growth of the isolated organ, was reported by White (1934) with excised tomato roots. Since then the excised roots of numerous herbaceous species have been cultured (Butcher & Street, 1964). Less success has been achieved in starting root cultures from woody plants, although there are reports on cultures from *Acacia* sp. (Bonner, 1942), *Robinia* sp. (Seeliger, 1956), *Acer rubrum* L. (Bachelard & Stowe, 1963), *Comptonia* sp. (Goforth & Torrey, 1977), and several species of gymnosperms (Brown & Sommer, 1975).

Some general comments can be made about the nutrition of cultured roots. In some cases, the minimum growth requirements were met with the essential mineral elements, a carbon source, a vitamin supplement, and a few amino acids. Some responded favorably to the addition of auxin and other growth regulators. Another effective growth stimulant for some isolated roots is *myo*-inositol, as it plays a role in secondary vascular tissue formation in excised roots of radish (*Raphanus*) (Loomis & Torrey, 1964; Torrey & Loomis, 1967). A marked improvement over White's medium is the substitution of a chelated form of iron for $Fe_2(PO_4)_3$ (Dalton, Iqbal, & Turner, 1983). NaFeEDTA, the sodium salt of ferric EDTA, is generally used as a chelated form of iron (see "Iron supplement" in Chapter 4). Sucrose is the carbon source of choice, although some monocot roots grow equally well with D-glucose. A sugar level of 1.5–2.0% (w/v) is sufficient, and higher concentrations may alter root metabolism (Guinn, 1963) (Figure 9.1). Although several amino acids have been tested, most of them inhibit the growth of cultured roots. The vitamin requirements vary slightly for different species, although all species require the addition of thiamine. The effect of light on the growth of cultured roots appears to vary with the species and the cultural conditions.

A question has been raised about the relationship between cultured roots and similar roots produced by an intact plant. Although isolated roots and "intact" roots are alike in many anatomical and metabolic ways, certain differences have been reported. Excised roots gradually lose the capability of forming secondary vascular tissues during culture. Cultured tomato roots, in contrast to seedling roots of the same plant, fail to show the normal geotropic response to gravity (Butcher & Street, 1964). In addition, the biochemical composition of cultured roots may differ from that of seedling roots (Abbott, 1963). Studies with microbial symbionts associated with roots have suggested that cultured roots may differ

Figure 9.1. Excised tomato (*Lycopersicon esculentum*) roots cultured 7 days at 27°C in White's medium containing sucrose at a concentration of (1) 0.5%, (2) 1.0%, (3) 1.5%, (4) 2.0%, (5) 3.0%, and (6) 4.0%. Note the formation of lateral roots from the main root axis. In order to subculture roots, the root is cut into sectors. Each sector, transferred to a fresh medium, contains a portion of the main root axis plus several lateral roots. (Courtesy of H. E. Street.)

in some respects from the roots of axenically grown seedlings (Torrey, 1978).

Rapidly growing hairy-root cultures, obtained by the genetic transformation of plant cells by *Agrobacterium rhizogenes*, may revolutionize the commercial production of rare chemicals (Rhodes et al., 1987). These roots are characterized by a high degree of lateral branching, a profusion of root hairs, and a stable and high-level biosynthesis of secondary metabolites. The transformed roots can be cultured indefinitely, with genetic stability, on a defined medium devoid of any growth regulators. *A. rhizogenes* can be a vector to introduce genes into hairy-root cultures. Biosynthesis genes of secondary metabolite formation can be engineered to be expressed in the cultured root and flanked by T-DNA border sequences in an expression plasmid. Inserted genes in the plasmid are cotransported to hairy roots along with the wild-type Ri T-DNA genes (Yamada & Hashimoto, 1990). A current project in England is investigating the bioengineering problems of large-scale hairy-root cultures with a 500-liter pilot plant bioreactor (Wilson et al., 1990). Large-scale cultures involving plant roots have been achieved in Japan. Cell aggregates with profuse proliferating roots of *Panax ginseng* have been cultured in a 20,000-liter bioreactor for saponin production by the Nitto Electric Co. (Yamada & Hashimoto, 1990).

The technique employed to initiate and subculture roots requires some explanation. Root tips about 10 mm in length are removed from young seedlings produced during the axenic germination of seeds, and these apical tips are transferred to an aqueous culture medium. After about a week the primary root produces lateral roots. The main root axis is subdivided into "sectors," each containing a portion of the main axis plus several lateral branch roots. Each of these sectors is transferred separately to a fresh medium for an additional period of growth. The lateral roots subsequently produce laterals themselves. Each of these sector cultures provides the investigator with a constant supply of lateral root tips for experiments, as well as a source of material for the propagation of the clone. This technique can be used only when the excised root produces laterals in some sequential order; in some species this does not occur.

In the present experiment sterile seedling roots obtained from the axenic germination of seeds taken from a fresh tomato fruit will be excised and cultured in a modified White's medium. After about 7–10 days of growth, sufficient lateral root development will have occurred to permit the first subculture with sector inocula (see Fig. 9.1).

Through periodic subcultures the student should maintain the root culture over a period of several weeks. Various experiments can be devised in order to determine the optimum conditions for growth and development of the isolated roots.

### LIST OF MATERIALS

*Sterilization mode:* C, chemical; O, oven; A, autoclave

- C  fresh ripe tomato (*Lycopersicon esculentum*) fruit with no surface blemishes or breaks in the skin
- O  stainless steel forceps
- C  scalpel; Bard–Parker No. 7 surgical knife handle with a No. 10 blade is recommended
- A  9-cm Petri dishes, each containing two sheets Whatman No. 1 filter paper (five)
- A  125-$cm^3$ Erlenmeyer flask containing 100 $cm^3$ $DDH_2O$ (one)
- A  10-$cm^3$ pipette; cotton plugged and enclosed completely in heavy wrapping paper
- A  125-$cm^3$ Erlenmeyer flasks, each containing 25 $cm^3$ modified White's medium (see "Formulations of tissue culture media" at the end of the book) (five)
- A  paper toweling; several sheets enclosed in envelope
- –  ethanol (80% v/v) dip
- –  ethanol (70% v/v) in plastic squeeze bottle
- –  methanol lamp
- –  graph paper (one sheet)
- –  incubator (25–27 °C)
- –  iridectomy scissors (optional)
- –  orbital platform shaker (optional)

### PROCEDURE

*Preparation of White's medium* (see "Formulations of tissue culture media" at the end of the book). In general, use the same procedure outlined in Chapter 4 for the preparation of the MS medium; that is, first prepare stock solutions of iron, micronutrients, and vitamins. Because of the minute amount of molybdenum required, weigh 10 mg $MoO_3$, dissolve it, and add $DDH_2O$ for a final volume of 1,000 cm$^3$. Pipette 1 cm$^3$ to deliver 0.01 mg $MoO^3$ to the micronutrient stock.

White's formulation required $Fe_2(SO_4)_3$. Because ferric sulfate precipitates easily from solution, a chelated form of iron is more desirable. The concentration of iron found in the MS medium can be used satisfactorily in White's medium. Prepare and use the stock as directed in Chapter 4.

The vitamins are dissolved in $DDH_2O$ in the order indicated, and a final stock volume of 100 cm$^3$ is prepared.

Add approximately 400 cm$^3$ $DDH_2O$ to a 1-liter beaker. Weigh and dissolve each of the macronutrient salts given in the table of formulations. Add by pipette to the macronutrient solution the following from each of the stock solutions: 5 cm$^3$ iron, 10 cm$^3$ micronutrients, and 1 cm$^3$ vitamins.

The final medium is adjusted to pH 5.5, and sucrose is supplied at a concentration of 20,000 mg/l. (*Note:* There is a possibility of some degradation of the B vitamins during autoclaving.)

## CULTURE PROCEDURE

The working area of the hood is lined with paper toweling, previously sterilized by autoclave. Wash the tomato fruit with tap water. Surface sterilize the skin of the fruit with ethanol (70% v/v) in the hood. Using sterile instruments, cut the skin and slice the fruit into four sectors. Pull the sectors apart exposing the seeds. Remove the seeds with the forceps and carefully remove the fruit pulp around each seed. Place approximately five seeds in each of the Petri dishes lined with filter paper. Moisten the filter paper with about 8–10 cm$^3$ sterile $DDH_2O$ delivered by pipette. Wrap the plates in plastic wrap and place them in a dark incubator (25–27 °C). The seeds will germinate within a few days. The dishes should be checked occasionally for signs of drying of the filter paper; if necessary, a few milliliters of sterile water may be added. Root tips about 10 mm in length are excised with a scalpel or iridectomy scissors and transferred to the flasks containing White's medium. White (1963) recommended that the root tips be transferred individually, that is, one per culture flask. After about one week, sufficient lateral root development will be evident, and sector inocula (see Fig. 9.1) can be removed and subcultured to a fresh medium. The student should perform weekly transfers and attempt to maintain the tomato root culture over a period of several weeks.

## RESULTS

Although cultured roots are difficult to measure in situ during culture, an attempt will be made to approximate the linear growth rate. Daily measurements can be made on a single culture, within the flask, without exposure and risk of contamination. Place the flask on a sheet of graph paper and align the root with the markings on the paper. Compare the growth rates of root cultures that have been stationary to the growth rates of other cultures that received agitation on a shaker. An interesting comparison can also be made on roots cultured in the dark versus others that received illumination. The incubation temperature will also influence the growth rate. The growth rates of tomato roots cultured under various conditions have been reported by Boll (1965).

## QUESTIONS FOR DISCUSSION

1. What are some advantages of using an aqueous medium in comparison to a gel-solidified medium? Can you think of any advantages that were not mentioned in this chapter?
2. In addition to aeration, what are some other possible effects of agitating cultured plant cells, cell aggregates, and organs?
3. What are the nutritional requirements for the culture of tomato roots?
4. What is the significance of hairy-root cultures? How are they produced?

## APPENDIX

*Salinity and root growth.* Repeat the experiment on the culture of tomato roots as outlined in this chapter except prepare a series of culture flasks numbered 1–6. Flask 1 contains White's medium as a control; flasks 2–6 contain White's medium supplemented with a range of concentrations of NaCl (0.1, 0.5, 1.0, 1.5, and 2.0% w/v). Approximate the linear growth rate, on a daily basis, of the tomato roots cultured in each of the flasks. Explain the results. How would you devise an experiment to determine whether the effect of NaCl on root growth was due to sodium toxicity or to osmotic potential?

*Initiation and culture of hairy roots.* Plants of tomato or tobacco are grown under greenhouse or growth room conditions until approximately 1.5 ft high. The stems of the plants are then wounded by inserting a sterile scalpel blade or dissecting needle. Immediately apply to the wound a small smear of *Agrobacterium rhizogenes* from a culture dish. (Smears of *A. rhizogenes* are available at cost from the American Type Culture Collection.) After approximately 10 days hairy roots will be growing out from the point of inoculation. Remove roots with a scalpel. Under a laminar flow hood, surface sterilize the roots in 1% hydrochlorite, rinse several times in sterilized distilled water, and transfer to White's culture medium as was described previously in this chapter.

### SELECTED REFERENCES

Abbott, A. J. (1963). The growth and development of excised roots in relation to trace element deficiencies. Ph.D. thesis, Dept. of Botany, University of Bristol.

Bachelard, E. P., & Stowe, B. B. (1963). Growth in vitro of roots to *Acer rubrum* L. and *Eucalyptus camaldulensis* Dehn. *Physiol. Plant.* 16, 20–30.

Boll, W. G. (1965). The growth curve and growth correlations of excised tomato roots grown in sterile culture. *Can. J. Bot.* 43, 287–304.

Bonner, J. (1942). Culture of isolated roots of *Acacia melanoxylon*. *Bull. Torrey Bot. Club* 69, 130–3.

Brown, C. L., & Sommer, H. E. (1975). *An atlas of gymnosperms cultured in vitro: 1924–1974.* Macon: Georgia Forest Research Council.

Butcher, D. N., & Street, H. E. (1964). Excised root cultures. *Bot. Rev.* 30, 513–86.

Dalton, C. C., Iqbal, K., & Turner, D. A. (1983). Iron phosphate precipitation in Murashige and Skoog media. *Physiol. Plant.* 57, 472–6.

Goforth, P. L., & Torrey, J. G. (1977). The development of isolated roots of *Comptonia peregrina* (Myricaceae) in culture. *Am. J. Bot.* 64, 476–82.

Guinn, G. (1963). Aseptic culture of excised roots: A limited review of recent literature. In *Plant tissue culture and morphogenesis*, ed. J. C. O'Kelley, pp. 35–45. New York: Scholar's Library.

Loomis, R. S., & Torrey, J. G. (1964). Chemical control of vascular cambium initiation in isolated radish roots. *Proc. Nat. Acad. Sci. USA* 52, 3–11.

Rhodes, M. J. C., Robins, R. J., Hamill, J. D., Parr, A. J., & Walton, N. J. (1987). Secondary product formation using *Agrobacterium rhizogenes*-transformed "hairy root" cultures. *Newsletter IAPTC 53*, 2–15.
Said, A. G. E., & Murashige, T. (1979). Continuous cultures of tomato and citron roots in vitro. *In Vitro 15*, 593–602.
Seeliger, T. (1956). Über die Kulture isolierter Wurzeln die Robinie (*Robinia pseudoacacia* L.). *Flora 144*, 47–83.
Street, H. E. (1957). Excised root culture. *Biol. Rev. 32*, 117–55.
  (1969). Growth in organized and unorganized systems: Knowledge gained by culture of organs and tissue explants. In *Plant physiology: A treatise*, ed. F. C. Steward, vol. VB, pp. 3–224. New York: Academic Press.
Torrey, J. G. (1965). Physiological bases of organization and development in the roots. In *Encyclopedia of plant physiology*, ed. W. Ruhland, vol. XV/1, pp. 1256–1327. Berlin: Springer-Verlag.
  (1978). In vitro methods in the study of symbiosis. In *Frontiers of plant tissue culture 1978*, ed. T. A. Thorpe, pp. 373–80. Calgary: IAPTC.
Torrey, J. G., & Loomis, R. S. (1967). Auxin–cytokinin control of secondary vascular tissue formation in isolated roots of *Raphanus*. *Am. J. Bot. 54*, 1098–1106.
White, P. R. (1934). Potentially unlimited growth of excised tomato root tips in a liquid medium. *Plant Physiol. 9*, 585–600.
  (1963). *The cultivation of animal and plant cells*, 2d Ed. New York: Ronald Press.
Wilson, P. D. G., Hilton, M. G., Meehan, P. T. H., Waspe, C. R., & Rhodes, M. J. C. (1990). The cultivation of transformed roots from laboratory to pilot plant. In *Progress in plant cellular and molecular biology*, ed. H. J. J. Nijkamp, L. H. W. van der Plas, & J. Van Aartrijk, pp. 700–5. Dordrecht: Kluwer Academic Publishers.
Yamada, Y., & Hashimoto, T. (1990). Possibilities for improving yields of secondary metabolites in plant cell cultures. In *Progress in plant cellular and molecular biology*, ed. H. J. J. Nijkamp, L. H. W. van der Plas, & J. Van Aartrijk, pp. 547–56. Dordrecht: Kluwer Academic Publishers.

# 10

## Micropropagation by bud proliferation

Micropropagation of ornamental and crop plants by the proliferation of shoot tips has resulted in the commercial production of plants on a worldwide scale. The entire method of micropropagation with shoot tips is based on the cytokinin-induced outgrowth of bud primordia, each of which produces a miniature shoot. Once a cluster of shoots has been formed, it can be subdivided into smaller clumps of offshoots, transferred to a fresh medium, and the process repeated. The rates of micropropagation vary greatly from species to species, but it is often possible to produce several million plants in the period of a year starting from a single isolated shoot tip (Thomas, 1981; Wilkins & Dodds, 1983).

    White (1933) made one of the first attempts to culture the shoot tip using *Stellaria media*. Later, Loo (1945) succeeded in culturing stem tips some 5 mm in length excised from *Asparagus officinalis*. The subculture of the cladophylls, formed on the cultured asparagus shoots, resulted in the formation of plantlets. From these early studies the concept was developed that ferns and angiosperms differ in the degree of autonomy of their isolated shoot apices. Isolated fern apices readily regenerated plants on a simple defined medium (Wetmore & Morel, 1949; Wetmore, 1953), whereas the cultural requirements for angiosperms appeared to be more complicated. Ball (1946) found that root initiation and plantlet formation occurred if the tip explant of angiosperms included a minimum of three leaf primordia and subjacent stem tissue. Later, Smith and Murashige (1970) succeeded in culturing apical meristems devoid of any leaf primordia from two species of *Nicotiana*, *Daucus carota* L., *Tropaeolum majus* L., and *Coleus blumei* Benth. The explants, consisting only of meristematic dome tissue, produced complete plants following culture on a defined medium.

Because of the extensive application of shoot-tip cultures in horticulture and plant pathology, there has been a flagrant misuse of the botanical nomenclature (Murashige, 1974). The terms "meristem culture," "meristemming," and "mericlones" have been used for the culture of relatively large stem tips; for example, orchid growers often use sections as long as 5–10 mm (Murashige, 1974). Cutter (1965) clearly distinguishes the apical meristem from the shoot apex. "Apical meristem" refers *only* to the region of the shoot apex lying distal to the youngest leaf primordium, whereas "shoot apex" refers to the apical meristem plus a few subjacent leaf primordia. The apical meristem explants employed by Smith and Murashige (1970) were about 80 μm in height. These minute explants are extremely difficult to excise, have a very low survival rate, and are impractical for the in vitro propagation of plants. Apical meristem cultures, however, are important in the development of pathogen-free stock.

The application of shoot-apex cultures for the clonal multiplication of plants was first realized by Morel (1960) during his studies on the propagation of the orchid *Cymbidium*, and modifications of his technique are currently used for the commercial production of orchids. The rapid progress in micropropagation techniques using multiple-shoot formation was due largely to the efforts of Murashige. Murashige (1974) subdivided the procedure into three stages:

I   establishment of the aseptic culture;
II  multiplication of propagula by repeated subcultures on a multiplication medium; and
III preparation of the plantlets for establishment in the soil.

Since Murashige's terminology, workers now recognize stages O and IV:

O  preparation of the mother plant prior to explant removal (Debergh & Maene, 1981; Debergh, 1987);
IV careful nurturing of the plantlets in a potting mix under in vivo conditions.

During stage O there are several practices for preparing the mother plant. For example, subirrigation to provide water and nutrition ensures lower humidity and cleaner explants. The regulation of temperature, daylength, and light intensity is important to ensure that the donor plant is in the proper physiological condition for explant removal. Treatment with either low temperature or $GA_3$ may be used to facilitate breaking dormancy for certain bulbs, tubers, or other dormant organs (Debergh, 1987).

Factors influencing stage I include the choice of a suitable explant, the composition of the medium, and the appropriate environmental conditions. Shoot tips and buds excised from healthy and actively growing herbaceous plants are generally ideal material for multiple-shoot production. The larger the tip explant, the more rapid the growth and greater the ability to survive (Hussey, 1983). A procedure that does not involve a callus phase is preferable, because the genetic instability of callus leads to a high degree of genetically aberrant plants. Although some groups of plants have unique nutritional needs, the MS formulation is satisfactory in most cases. The basal medium is supplemented with vitamins, sucrose, and the appropriate growth regulators. The cultures can be grown on agar or on filter-paper bridges (Goodwin, 1966) over a liquid medium. Light, necessary for photomorphogenesis and chlorophyll biosynthesis, is provided with Gro-Lux and white fluorescent tubes. Murashige (1974) found that many cultures grew best with 1,000 lux for stages I and II, with the light intensity increased to 3,000–10,000 lux for stage III. Photoperiods of 16 hr were optimum for several species (Murashige, 1977), although relatively little research has been done on the effects of light and temperature on micropropagation.

Typically, the same medium and environmental conditions are used for both stages I and II. The choice and concentration of growth regulators are the most important consideration in preparation of the medium. Cytokinins may be added in the form of kinetin, benzylaminopurine (BAP), 2iP, or zeatin. Some of Murashige's multiplication media that are commercially available contain combinations of adenine sulfate plus either kinetin or IPA. The source of exogenous auxin is usually IAA, $\alpha$-NAA, or IBA (indole-3-butyric acid). The auxin 2,4-D is unsatisfactory, since it stimulates callus formation and suppresses organogenesis. Gibberellic acid ($GA_3$) may be required for the culture of some shoot apices. Morel (1975) reported that isolated apices of potato, *Dahlia*, carnation, and *Chrysanthemum* need exogenous $GA_3$ (0.1 mg/l) for the normal development of the shoot apex and subsequent micropropagation.

Stage III involves the development of a root system, hardening the young plants to moisture stress, increasing resistance to certain pathogens, and conversion of the plants to an autotropic state (Murashige, 1974). Root initiation may be facilitated by adding a low concentration of either $\alpha$-NAA or IBA to the medium. The auxin treatment must be limited to a brief period of time. Auxin at this stage of the process may have undesirable side effects – that is, may stimulate callus production and inhibit root elongation (Hu & Wang, 1983). The formation of roots often occurs after transfer to a medium lacking hormones. In fact, the shoots of some species can be rooted by conventional root procedures after removal from the in vitro environment (Debergh & Maene, 1981).

During stage IV the plantlets are transferred from the culture tubes to a soil mixture. This should be a moist sterile commercial product, such as Cole's Starting Mix for Seeds and Cuttings. The plantlets must undergo a period of acclimation to in vivo environmental conditions. They must be protected from direct sunlight, and the relative humidity should be gradually decreased over a period of time. During the acclimation period the rate of photosynthesis is initially low, and survival of the plantlets may depend on the accumulation of carbohydrates during the culture (Torres, 1989). Another difficulty is the relatively thin layer of cuticular wax present, which may lead to a serious dehydration of the aerial tissues (Sutter & Langhans, 1979).

Although the micropropagation of many tree crop species has been achieved using shoot-apex cultures, woody plants pose some unusual problems. Bud cultures must be taken either from shoots in the juvenile growth phase or selected from rejuvenated shoots. Buds taken from mature trees in the adult phase have little capacity for micropropagation since the cultured material is incapable of producing roots (Pierik, 1990). "Rejuvenation," meaning a return to the juvenile state, involves a complete reversal of maturation either by sexual reproduction or, in some cases, vegetative propagation (Pierik, 1990). The term "reinvigoration" refers to a reversal of aging. For example, increased rooting capability results from pruning, hedging, BAP treatment, and other devices. So far it is impossible to distinguish clearly between rejuvenation and reinvigoration since the underlying biochemistry of the processes is obscure. Pierik (1990) has reviewed the status of the problems involved.

Shoot-apex cultures of woody plants require several consecutive treatments (Wilkins & Dodds, 1983). The excised bud, with the formation of a rosette of leaves, requires exogenous gibberellin and cytokinin for growth. The explant must then be transferred to another medium lacking exogenous growth regulators in order to promote stem elongation. Finally, the culture must be transferred to a third medium containing exogenous auxin for the initiation of roots (Morel, 1975).

Tissues containing relatively high concentration of phenolic compounds are difficult to culture. Polyphenolases stimulated by tissue injury will oxidize these phenolics to growth-inhibiting, dark-colored compounds. Techniques used to suppress this metabolic sequence include the following (Hu & Wang, 1983):

1. adding antioxidants to the medium,
2. presoaking the explants in antioxidant solutions prior to culture,
3. subculturing to a fresh medium on signs of enzymatic browning, and
4. providing little or no light during the initial period of culture.

For example, the brown exudate resulting from polyphenolase activity in *Eucalyptus grandis* cultures was eliminated by the following protocol. Explants were prepared from young stem tissues and presoaked in sterile distilled water for 3 hr prior to culture. The explants, leached of phenolic compounds, were then initially incubated in darkness for several days (Durand-Cresswell, Boulay, & Francelet, 1982). Additional information on the elimination of polyphenolase activity can be found in the review by Hu and Wang (1983).

In the following experiment the student will excise and micropropagate some shoot apices from either potato (*Solanum tuberosum*) or geranium (*Pelargonium*). An attempt will be made to propagate whole plants from the excised and cultured shoot apices. The explants are cultured individually in a rimless Pyrex test tube. The excised shoot apex is transferred to the surface of an agar-solidified MS basal salt medium with 0.8% agar and 2% sucrose.

## LIST OF MATERIALS

*Sterilization mode:* C, chemical; O, oven; A, autoclave

-   shoot-apical material from potato (*Solanum tuberosum*) or geranium (*Pelargonium*)
- C  scalpel or razor blade for microsurgery; either use a surgeon's scalpel fitted with a No. 11 blade (Shabde-Moses & Murashige, 1979) or prepare microscalpels by flattening the tips of No. 2 insect mounting pins (Smith & Murashige, 1970)
- A  Pyrex test tubes (16 × 150 mm), each containing approximately 5 cm$^3$ of MS medium supplemented with IBA (2.0 mg/l), BAP (2.0 mg/l), GA$_3$ (0.1 mg/l), sucrose (2.0 w/v), and agar (0.8%); close with foil caps (12)
- A  9-cm Petri dishes, each containing two sheets Whatman No. 50 filter paper (five)
- A  125-cm$^3$ Erlenmeyer flasks, each containing 100 cm$^3$ DDH$_2$O (eight)
- A  5-cm$^3$ pipette
- O  forceps
- O  250-cm$^3$ beakers (four)
-   100 cm$^3$ aqueous 10% (v/v) solution commercial bleach
-   ethanol (80% v/v) dip
-   ethanol (70% v/v) in plastic squeeze bottle
-   methanol lamp
-   binocular dissection microscope with external top lighting (50×)
-   plant growth chamber (16/8 hr photoperiods; 1,000 lux illumination; 27 °C; 70% relative humidity

## PROCEDURE

1. Add by pipette 5 cm$^3$ of MS medium (supplemented as indicated) to each of the culture tubes. The preparation of the MS medium is given in Chapter 4. Cap each tube with aluminum foil. The entire assembly is sterilized by autoclave and permitted to cool.

2. Apical shoots 1.0 cm in length are removed from the plant and placed in the hypochlorite solution for approximately 15 min. All subsequent procedures are conducted aseptically.

3. After surface sterilization, rinse the shoots several times in sterile $DDH_2O$. Transfer the shoots to a Petri dish containing Whatman No. 50 filter paper for surgical removal of the shoot apex. Moisten the filter paper with a few cubic centimeters of sterile $DDH_2O$ to prevent desiccation of the shoot tips.

4. A binocular dissection microscope is necessary for viewing the shoot apex; a magnification of 50× is adequate. The terminal 0.5 mm of the tip is carefully excised with a scalpel fitted with a pointed tip blade (No. 11). The excision should yield an explant consisting of the apical dome plus a small number of leaf primordia. In addition, several shoot tips should be removed varying in length from 1 to 5 mm. Observations will be made on the optimal size of the explant for purposes of micropropagation.

5. Each shoot tip is carefully transferred with the blade of the scalpel, or some type of microscalpel, to the surface of the agar-solidified culture medium.

6. The culture tubes, capped with foil, are cultured in a plant growth chamber set for 16/8 hr photoperiods at 27 °C and relative humidity of about 70%. An intensity of 1,000 lux is sufficient because the explants should not be exposed to high intensities of light (Shabde-Moses & Murashige, 1979).

## RESULTS

The time required for the excised shoot tips to initiate growth and begin micropropagation, that is, with the outgrowth of axillary buds, may vary from a few weeks to several months depending on the species and the cultural conditions.

Potato is one of the fastest-growing plants in micropropagation. Within five or six weeks a dense cluster of proliferating shoots should be evident (Fig. 10.1). After the shoots are well developed they are excised and transferred to a root-induction medium, which is the same as that used in this experiment except for the omission of BAP. Root initiation normally takes four to six weeks.

Once an adequate root system has developed, the young plantlets are ready to be transferred to nonsterile condition. The plantlets are first placed in a sterile soil mixture and maintained under humid conditions by mist irrigation. This is preferable to sealing the pots with plastic, which tends to encourage the growth of fungi. Gradually the young plants are hardened off by reducing the mist irrigation, and eventually they are transferred to a cool greenhouse. Finally, the plants regenerated from the shoot-apex cultures are planted in the field.

Figure 10.1. Micropropagation of potato (*Solanum tuberosum*). (a) Excised shoot tip grows to produce a plantlet. (b) In micropropagating culture multiple shoots are produced by outgrowth of axillary buds. (c) In vitro plantlet transferred to "jiffy" pot. (d) Plantlet derived from shoot tip ready for transfer to field.

## QUESTIONS FOR DISCUSSION

1. Briefly describe the difference between the apical meristem and the shoot apex. Why are shoot apices preferred for the propagation of plants?
2. What are some problems associated with the growth and development of cultured shoot apices excised from woody plants?

## SELECTED REFERENCES

Ball, E. (1946). Development in sterile culture of stem tips and subjacent regions of *Tropaeolum majus* L. and *Lupinus albus* L. *Am. J. Bot. 33*, 301–18.

Cutter, E. G. (1965). Recent experimental studies of the shoot apex and shoot morphogenesis. *Bot. Rev. 31*, 7–113.

Debergh, P. C. (1987). Improving micropropagation. *IAPTC Newsl. 51*, 2–10.

Debergh, P. C., & Maene, L. J. (1981). A scheme for commercial propagation of ornamental plants by tissue culture. *Sci. Hortic. 14*, 335–45.

Durand-Cresswell, R., Boulay, M., & Francelet, A. (1982). Vegetative propagation of *Eucalyptus*. In *Tissue culture in forestry*, ed. J. M. Bonga & D. J. Durzan, pp. 150–81. The Hague: Martinus Nijhoff/Junk.

Goodwin, P. B. (1966). An improved medium for the rapid growth of isolated potato buds. *J. Exp. Bot. 17*, 590–5.

Hu, C. Y. & Wang, P. J. (1983). *Meristem, shoot tip, and bud cultures*. In *Handbook of plant cell culture*, vol. 1, *Techniques for propagation and breeding*, ed. D. A. Evans, W. R. Sharp, P. V. Ammirato, & Y. Yamada, pp. 177–227. New York: Macmillan.

Hussey, G. (1983). In vitro propagation of horticultural and agricultural crops. In *Plant biotechnology*, ed. S. H. Mantell & H. Smith, pp. 111–38. Cambridge: Cambridge University Press.

Loo, S. W. (1945). Cultivation of excised stem-tips of asparagus in vitro. *Am. J. Bot. 32*, 13–17.

Morel, G. (1960). Producing virus-free cymbidiums. *Am. Orchid Soc. Bull. 29*, 495–7.

(1975). Meristem culture techniques for the long-term storage of cultivated plants. In *Crop genetic resources for today and tomorrow*, ed. O. H. Frankel & J. G. Hawkes, pp. 327–32. Cambridge: Cambridge University Press.

Murashige, T. (1974). Plant propagation through tissue culture. *Annu. Rev. Plant Physiol. 25*, 135–66.

(1977). Manipulation of organ culture in plant tissue culture. *Bot. Bull. Acad. Sin. (Taipei) 18*, 1–24.

Pierik, R. L. M. (1990). Rejuvenation and micropropagation. In *Progress in plant cellular and molecular biology*, ed. H. J. J. Nijkamp, L. H. W. van der Plas, & J. Van Aartrijk, pp. 91–101. Dordrecht: Kluwer Academic Publishers.

Shabde-Moses, M., & Murashige, T. (1979). Organ culture. In *Nicotiana: Procedures for experimental use*, ed. R. D. Durbin, pp. 40–51. Washington, D.C.: USDA Tech. Bull. No. 1586.

Smith, R. H., & Murashige, T. (1970). In vitro development of the isolated shoot apical meristem of angiosperms. *Am. J. Bot. 57*, 562–8.

Sutter, E., & Langhans, R. W. (1979). Epicuticular wax formation on carnation plantlets regenerated from shoot tip cultures. *J. Am. Soc. Hortic. Sci. 104*, 493–6.

Thomas, E. (1981). Plant regeneration from shoot culture derived protoplasts of tetraploid potato (*Solanum tuberosum*). *Plant Sci. Lett. 23*, 81–8.

Torres, K. C. (1989). *Tissue culture techniques for horticultural crops*. New York: Van Nostrand Reinhold.

Wetmore, R. H. (1953). The use of in vitro cultures in the investigations of growth and differentiation in vascular plants. *Brookhaven Symp. Biol. 6*, 22–40.

Wetmore, R. H., & Morel, G. (1949). Growth and development of *Adiantum pedatum* L. on nutrient agar. *Am. J. Bot. 36*, 805–6.

White, P. R. (1933). Plant tissue culture: Results of preliminary experiments on the culturing of isolated stem-tips of *Stellaria media*. *Protoplasma 19*, 97–116.

Wilkins, C. P., & Dodds, J. H. (1983). Tissue culture propagation of temperate fruit trees. In *Tissue culture of trees*, ed. J. H. Dodds, pp. 56–77. London: Croom Helm.

# 11

## Anther and pollen cultures

The cells of haploid plants contain a single complete set of chromosomes, and these plants are useful in plant-breeding programs for the selection of desirable characteristics. The phenotype is the expression of single-copy genetic information, there being no masking of a trait through gene dominance. The purpose of anther and pollen culture is to produce haploid plants by the induction of embryogenesis from repeated divisions of monoploid spores, either microspores or immature pollen grains. "Microspores" represent the beginning of the male gametophyte generation; "pollen grains" are mature microspores, especially following their release from tetrads (Bhojwani & Bhatnagar, 1974). The chromosome complement of these haploids can be doubled by colchicine or by regeneration techniques to yield fertile homozygous diploids (Vasil & Nitsch, 1975). Although the number of successful pollen culture systems is still relatively small (Sunderland, 1971, 1977), this technique has resulted in several improved varieties of crop plants in China (Sunderland & Cocking, 1978; Sunderland, 1979).

Tulecke (1953) first observed that mature pollen grains of the gymnosperm *Ginkgo biloba* could be induced to form a haploid callus following culture on a suitable medium. Repeated divisions of cultured pollen grains of angiosperms were described much later by Guha and Maheshwari (1966), who made a remarkable discovery by accident. These investigators were conducting experiments with cultured pollen grains of *Datura innoxia* in order to determine the feasibility of this system for the study of factors regulating meiosis. The growth response of the pollen grains, enclosed in mature anthers, was of three types and reflected the nature of the medium. Although the pollen grains were unresponsive in the presence of IAA, callus was initiated on media containing either yeast extract or casein hydrolysate. Torpedo-shaped embryoids, which later developed into plantlets, were produced following culture of the anthers on media containing either kinetin or coconut water. Acetocarmine staining revealed that these newly formed plantlets contained only a single set of chromosomes (Guha & Maheshwari, 1966, 1967).

The particular stage of development of the anthers at the time of culture is the most important factor in achieving success in the formation of embryoids. In angiosperms with an indeterminate number of anthers in each flower bud, buds can be selected that will contain several anthers in various stages of pollen development. In species with a determinate number of anthers per flower, a series of buds must be examined in order to give all the stages of development. Two basic methods are used:

1. excised anthers are cultured on an agar or liquid medium, and embryogenesis occurs within the anther; or
2. the pollen is removed from the anther, either by mechanical means or by natural dehiscence of the anther, and the isolated pollen is cultured on a liquid medium.

It may take three to eight weeks for haploid plantlets to emerge from the cultured anthers (Reinert & Bajaj, 1977).

Sunderland (1979) reported that in flowers of many plants anthers fall into one of three categories: premitotic, mitotic, or postmitotic. In the premitotic category the best response is obtained by using anthers in which the microspores have completed meiosis but have not yet started the first pollen division (e.g., *Hyoscyamus, Hordeum vulgare*). Anthers of plants belonging to the mitotic group respond optimally about the time of the first pollen division (e.g., *Nicotiana tabacum, Datura innoxia, Paeonia*). The early bicellular stage of pollen development is best in the postmitotic plants (e.g., *Atropa belladonna, Nicotiana* spp.). In the case of *N. tabacum*, floral buds with the corolla barely visible beyond the calyx will probably contain anthers at the appropriate stage of development, although there may be slight differences among different cultivars (Kasperbauer & Wilson, 1979). This particular stage has been termed "stage-2" by Nitsch (1969) and "stage-4" by Sunderland and Dunwell (1977).

As discussed previously in Chapters 4 and 8, activated charcoal has a stimulatory effect on somatic embryogenesis as well as on the initiation of embryos from haploid anther tissue. This charcoal effect has been demonstrated for anther cultures of tobacco (Anagnostakis, 1974; Bajaj, Reinert, & Heberle, 1977), rye (Wenzel, Hoffmann, & Thomas, 1977), potato (Sopory, Jacobsen, & Wenzel, 1978), and other plants. The removal of inhibitory substances from the agar is considered to be a factor since a similar response was obtained by dialyzing agar against activated charcoal, or by employing highly purified agar (Kohlenbach & Wernicke, 1978). Another possibility is adsorption by charcoal of 5(hydroxymethyl)-2-furfural, a degradation product of autoclaved sucrose (Weatherhead, Bordon, & Henshaw, 1978). A study of anther cultures of *Petunia* and *Nicotiana* was consistent with the hypothesis that charcoal removes both endogenous and exogenous growth regulators from the culture medium (Martineau, Hanson, & Ausubel, 1981). Although the precise role of activated charcoal in this developmental process remains unknown, the use of charcoal for the enhancement of haploid plantlet production should be encouraged (Bajaj, 1983).

Another factor in anther culture is the physiological status of the parent plant (e.g., photoperiod, light intensity, temperature, and mineral nutrition). Anthers should be taken from flowers produced during the beginning of the flowering period of the plant (Sunderland, 1971). Higher yields of embryos have been reported from donor plants grown under short days and high light intensities (Dunwell, 1976). For additional information see Sunderland and Dunwell (1977).

Various types of anther pretreatment have been found to improve embryo production in some plants. Low-temperature pretreatment of anthers for periods of 2–30 days at temperatures of 3–10 °C may stimulate embryogenesis (Sunderland & Roberts, 1977). Other types of pretreatment include soaking the detached inflorescence in water for several days (Wilson, Mix, & Foroughi-Wehr, 1978) and centrifugation of the anthers at 3–5 °C for approximately 30 min (Sangwan-Norreel, 1977).

The presence of anther tissue in the culture introduces several ill-defined factors that influence embryoid production. Raghavan (1978) suggested that a gradient of endogenous auxin within the anther may play a role in pollen grain development. Embryogenic pollen grains were observed to be confined to the periphery of the anther locule in close proximity to the tapetum, and possibly substances released from the tapetum initiate embryogenic divisions in pollen grains within cultured anther segments of *Hyoscyamus niger* (Raghavan, 1978). Pollen is also sensitive to toxic substances released following injury to anther wall tissue (Horner & Street, 1978).

It is important to determine the chromosome number of the newly formed plantlets because there may be considerable variation in ploidy levels, depending on the developmental events that led to embryoid formation. Diploid heterozygous plants may arise from anther tissue or from growth of the microspore mother cells and unreduced microspores. The further development of dyads and incomplete tetrads often produces plants that are heterozygous at certain loci because of crossing-over prior to the first reduction division. Chromosome doubling and fusion of nuclei can produce homozygotes with varying ploidy levels. Plantlets formed from the callus tissue arising from haploid microspores can exhibit mutations and chimeras (Thomas & Davey, 1975). It was found that the ploidy levels of 2,496 rice plants derived from pollen cultures were 35.3% haploid, 53.4% diploid, 5.2% polyploid, and 6.0% mixoploid (Hu Han et al., 1978). Some developmental pathways exhibited by microspores are shown in Figure 11.1.

There are several techniques for doubling the chromosome number of the haploid plants, and two approaches can be taken:

1. regeneration by tissue culture methods, and
2. chemically induced doubling with colchicine.

The ploidy level of the plant involved must first be confirmed with standard cytological procedures before additional experiments are undertaken (Darlington & LaCour, 1976; Collins, 1979). One method of chromosome doubling employs aged leaf tissue from haploid plants because older leaves have the potential to regenerate both haploid and diploid plants. The diploids result from chromosome endoreduplication, which frequently occurs in cultured plant tissues. A technique for inducing plantlet formation starting with leaf explants from haploids has been given by Kasperbauer and Wilson (1979). In addition, the chromosome number can be doubled by the application of colchicine to either the embryos or the haploid plants. A simple procedure is to immerse the anthers containing the newly formed plantlets in an aqueous solution of colchicine (0.5% w/v) for 24–48 hr (Burk, Gwynn, & Chaplin, 1972). Another approach is to apply a preparation of colchicine in lanolin paste (0.4% w/v) to the axillary buds of decapitated mature haploid plants (Tanaka & Nakata, 1969). An evaluation of the technique for chromosomes is given by Jensen (1974).

Figure 11.1. Some possible developmental pathways of microspores under in vivo and in vitro conditions. Normal development (in vivo) results in the production of two sperm, a tube nucleus or cell, and pollen tube formation (*far left*). Several possibilities exist following the in vitro culture of isolated microspores or anthers. Either the tube or generative nucleus degenerates, and the surviving nucleus divides repeatedly and ultimately produces a haploid embryoid. The first mitosis of the microspore may produce two similar nuclei, and repeated division of these nuclei can produce a haploid embryoid. The similar nuclei, however, may fuse and produce a diploid embryoid (*far right*). Haploid callus of microspore origin can form embryoids de novo. (Diagram modified from Devreux, 1970.)

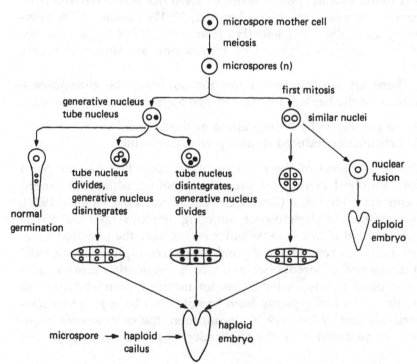

The potential for using haploid plants and homozygous lines in plant-breeding programs has long been recognized. One important area of research concerns the development of homozygous lines for the production of hybrids in self-incompatible species (e.g., rye and rape) (Keller & Stringham, 1978). Microspore culture is important in mutagenic studies: Mutations are not masked in haploids because there cannot be a dominant gene. For these studies to be successful, however, large numbers of microspores must be induced to undergo embryogenesis and develop on nutrient media. The haploids must remain genetically stable, and it must be possible to regenerate diploid plants from the haploids. This has been achieved in the case of *Nicotiana tabacum* (Devreux & Laneri, 1974).

Three different techniques are introduced in this chapter. In the first experiment the student will culture anthers of either henbane (*Hyoscyamus niger*) or tobacco (*Nicotiana tabacum*), both of which are in the Solanaceae family. An attempt will be made in the second experiment to induce the formation of haploid plantlets via the culture of isolated pollen removed from excised anthers. Assuming success in the previous experiments, the student will then attempt to obtain chromosome doubling and the production of diploid plants from the haploids.

## LIST OF MATERIALS

*Sterilization mode*: C, chemical; O, oven; A, autoclave

-   plants of *Hyoscyamus niger* or *Nicotiana tabacum* in the flower bud stage of development
- C  scalpel, with narrow blade (Bard–Parker No. 11)
- O  forceps (two)
- O  dissecting needles prepared from soft glass rod
- O  9-cm Petri dishes (10)
- O  culture tubes required for Experiment 1 (10)
- O  5-$cm^3$ pipettes
- O  Pasteur pipettes
- A  125-$cm^3$ Erlenmeyer flasks, each containing 100 $cm^3$ $DDH_2O$ (five)
- A  100 $cm^3$ MS medium supplemented with sucrose (2% w/v) and agar (0.6–0.8% w/v) required for Experiment 1
- A  50 $cm^3$ liquid MS medium supplemented with sucrose (2% w/v) required for Experiment 2

- 100 cm$^3$ aqueous solution (10% v/v) commercial bleach containing 2 cm$^3$ detergent or wetting agent (1% v/v)
- ethanol (80% v/v) dip
- ethanol (70% v/v) in plastic squeeze bottle
- methanol lamp
- dissection microscope
- microscope slides; cover slips
- acetocarmine stain
- Fuchs–Rosenthal hemocytometer, 0.2 mm in depth
- incubator or refrigeration unit for chilling buds (preset 7–8 °C)
- growth chambers equipped with fluorescent lighting (300 lux; 5,000 lux)
- root-induction medium required for Experiment 1 (after four to five weeks)
- colchicine (0.5% w/v)
- Parafilm (one roll)

PROCEDURES

*Experiment 1. Anther culture*

1. Plants selected for experimentation are cultivated until they reach the flower bud stage. In the case of tobacco plants, the buds should have a corolla length of 21–23 mm, and at this stage the pollen will have completed the first mitosis (Sunderland & Roberts, 1977). The methods used for isolation and culture are shown in Figure 11.2.

2. If tobacco is employed, best results will be obtained by chilling the buds approximately 12 days (7–8 °C) prior to culture (Sunderland & Roberts, 1977). For surface sterilization the buds are transferred to a Petri dish containing hypochlorite solution plus a wetting agent (10 min).

3. Rinse the buds several times in sterile DDH$_2$O. Using forceps and a dissecting needle, carefully tease open the buds and remove the anthers. Great care is required and a dissection microscope is necessary for this step. The dissected anthers from each bud are grouped together as they are removed.

4. One anther from each group is removed and squashed in acetocarmine in order to determine the stage of pollen development. If the pollen in the squashed anther is in the correct stage of

development, then the remaining anthers from that bud are placed into culture. In the case of *Hyoscyamus niger*, the microspores should have completed meiosis but not yet initiated the first pollen division. The pollen of *Nicotiana tabacum* should exhibit the first pollen division. The filaments must be removed before culture or they will form callus at the cut ends.

5. Anthers can be cultured either on agar-solidified culture media (Sunderland & Wicks, 1971) or by floating them on the surface of a liquid medium (Wernicke & Kohlenbach, 1976, 1977; Sunderland & Roberts, 1977). Directions for the preparation of the culture medium are given at the end of this section.

6. The anthers are cultured at 25 °C in light or darkness. After plantlet formation has been initiated, light is essential for the production of chlorophyll and normal plant growth. Continuous illumination from cool-white fluorescent lamps (300 lux) gives satis-

Figure 11.2. Basic procedure for the production of haploid plants from anther culture. Isolated buds are surface sterilized and the anthers aseptically removed. Individual anthers are screened by an acetocarmine staining method for the selection of the proper stage of pollen development. Subsequent culture results in embryoid and haploid plantlet formation.

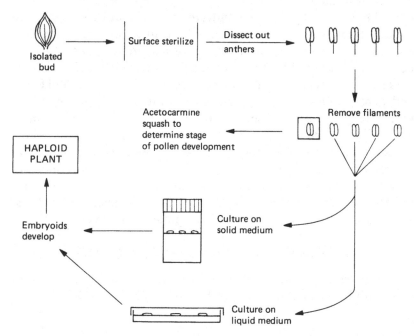

factory results (Kasperbauer & Wilson, 1979). Young embryos can be detected by gently bursting the cultured anthers on a microscope slide in a drop of acetocarmine stain. Plantlet formation occurs after a period of approximately four to five weeks of culture.

7. Separate the plantlets with forceps and discard the remaining anther tissue. In order to hasten the development of the plantlets, transfer them to a root-inducing medium after they are about 3 mm in length. This medium is identical to the anther culture medium except that the agar is reduced to 0.5% (w/v) and all other components are provided in half strength. During this period of growth the plants should be placed on a 12-hr daylength provided with 5,000-lux illumination from white fluorescent lamps (Kasperbauer & Wilson, 1979). It is recommended that the student examine the root tips of the plantlets with acetocarmine or Feulgen staining in order to verify the haploid chromosome number (Darlington & LaCour, 1976; Collins, 1979).

*Experiment 2. Pollen culture (Sunderland & Roberts, 1977)*

1–4. Same as for Experiment 1.

5. This technique is based on the release of pollen into the culture medium following the dehiscence of the tobacco anthers. Pretreatment of the tobacco buds by chilling (step 2) apparently facilitates the dehiscence. For each culture place the anthers from three tobacco buds in 5 $cm^3$ of liquid medium in a Petri dish. Remove and discard the anthers from some cultures after 6, 10, and 14 days. For anthers other than those of tobacco, it may be advisable to make a slight incision in the anther tissue with a fine scalpel blade and gently squeeze the contents of the anthers into the medium with forceps. Seal the dishes with Parafilm and incubate them at 28 °C in the dark for the first 14 days of culture. After 14 days transfer the cultures to an illuminated growth chamber (GroLux fluorescent lamps, 500 lux, 12-hr daylength, 25 °C). The pollen released from the anthers into the medium at 6-, 10-, and 14-day intervals will develop into haploid embryos (Sunderland & Roberts, 1977; Kasperbauer & Wilson, 1979; Nagmani & Raghavan, 1983).

*Experiment 3. Regeneration of diploid plants*

In this experiment an attempt will be made to promote chromosome doubling of the plantlets formed from the anther culture experiment. When plantlets are beginning to emerge from the cultured anthers, immerse the anthers in an aqueous solution of colchicine (0.5% w/v) for 24–48 hr. Colchicine is a powerful poison, and students should exercise the utmost care in handling this alkaloid (Appendix to Chapter 7). Following the colchicine treatment, rinse the plantlets in DDH$_2$O and culture them as described in Experiment 1, step 7. It is important to ascertain the chromosome number of the plantlets with acetocarmine or Feulgen staining. If diploidy is not achieved, the immersion time may be increased to 96 hr, and it may be necessary to repeat the treatment several times in order to achieve success (Sunderland & Dunwell, 1977).

## Preparation of anther and pollen culture media

The anthers of many plants, including *Nicotiana tabacum* and *Hyoscyamus niger*, can be cultured on relatively simple media consisting of minerals and sucrose. Other species, however, require the addition of organic supplements and plant hormones. Unfortunately, the addition of auxin to the culture medium appears to stimulate callus formation (Raghavan, 1978). The medium recommended for the culture of anthers in Experiment 1 is the MS medium supplemented with 2% (w/v) sucrose. If desired, the medium may be solidified with agar (0.6–0.8% w/v).

It is suggested that the student employ an alternative medium, devised by Sunderland and Roberts (1977), for the liquid culture of isolated pollen. Its composition is given in Table 11.1.

Table 11.1. *A basal medium for the liquid culture of pollen*

| Constituent | Concentration (mg/l) |
|---|---|
| KNO$_3$ | 950 |
| NH$_4$NO$_3$ | 825 |
| MgSO$_4$·7H$_2$O | 185 |
| CaCl$_2$ | 220 |
| KH$_2$PO$_4$ | 85 |
| FeSO$_4$·7H$_2$O | 27.8 |
| NA$_2$EDTA | 37.3 |
| sucrose | 20,000 |

## Preparation of acetocarmine stain

Acetocarmine is prepared by refluxing an aqueous solution of carmine (4% w/v) for 24 hr in acetic acid (50% v/v), and then filtering the product. Species with a low DNA content should first be stained with the Feulgen reaction. Anthers are fixed in acetic acid: ethanol (1:3 v/v) for several hours (4 °C), passed through a graded ethanol series to water, and hydrolyzed in HC1 (5 N) for 1 hr at room temperature. The root tips or spores may be stained with the Feulgen reagent for approximately 2 hr, and then squashed in acetocarmine (Sunderland & Dunwell, 1977).

## RESULTS

Remove the anthers from the culture medium at regular intervals and observe microscopically the development of the pollen grains. For ease of interpretation, Figure 11.3 shows the development of embryoids from *Hyoscyamus niger* pollen as viewed with the scanning electron microscope (Dodds & Reynolds, 1980). After two or three days in culture certain developmental changes can be detected (Fig. 11.3). Many of the pollen grains accumulate starch granules, others degenerate, and a small proportion divide and enter into the developmental pathway leading to the formation of haploid embryoids (Raghavan, 1975, 1978). The stages of embryoid development arising from cultured pollen grains are similar to the stages of zygotic embryo development normally found in diploid plants. The embryogenic pollen grains eventually rupture, and the developing system undergoes repeated cell divisions. This cellular proliferation gives rise to a globular-stage

Figure 11.3 (*facing*). Observations on embryoid development in cultured anthers of *Hyoscyamus niger* (henbane) with scanning electron microscopy. (a) Developing pollen grain after culturing for 20 hr. (b) Swollen grain beginning to break open at raphe after 5 days of culture. (c) Globular stage of embryoid after 7 days of culture showing association with the parental pollen grain. (d) Another embryoid showing globular stage after 7 days of culture. (e) Heart-shaped embryoid after 9 days of culture. (f) Embryoid bursting through the anther wall after 14 days of culture; note the developing roots and shoots. (From Dodds & Reynolds, 1980.)

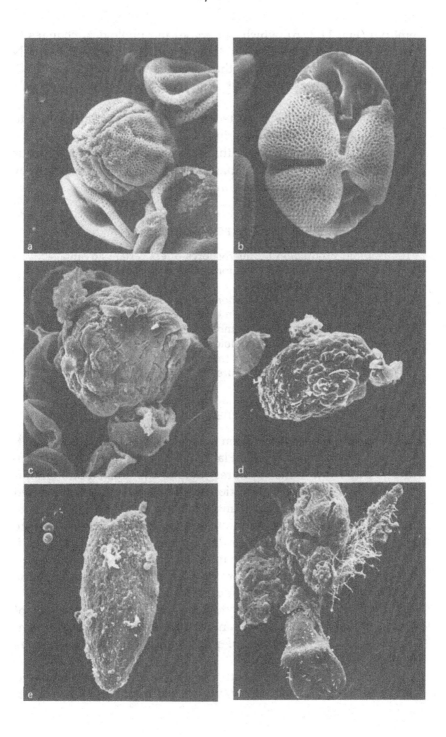

embryo (Fig. 11.3c,d), which undergoes morphological changes to produce heart-shaped (Fig. 11.3e) and torpedo forms. Eventually the haploid embryoids break through the anther wall (Fig. 11.3f). This entire process takes about 14 days in *Hyoscyamus*, and 21–28 days in *Nicotiana*.

The plantlets can be dissected from the anthers and grown to maturity. As outlined in Experiment 3, it is possible to treat these haploid plants in such a way that they develop into homozygous and fertile diploid plants. Haploid plants are sterile because they are unable to undergo reduction division in meiosis.

## QUESTIONS FOR DISCUSSION

1. Haploid production by microspores offers certain advantages over other methods for the formation of haploid plants. Comment on this statement.
2. Offer some possible explanations why cultured anthers will permit pollen to develop into embryos, whereas cultured isolated pollen grains may not form embryos.
3. Of what importance are haploid plants to the plant breeder?
4. What are the three pathways that *Hyoscyamus* pollen grains may follow when placed into culture?

## APPENDIX

1. Try screening anthers from several species of plants in an attempt to produce embryoids. Plant families that have yielded haploid plants from anther cultures include the Solanaceae, Gramineae, and Cruciferae. In addition to the plant species mentioned in this chapter, some other plants that may be useful include *Populus*, tomato (*Lycopersicon esculentum*), *Petunia hybrida*, maize (*Zea mays*), sugar beet (*Beta vulgaris*), and wheat (*Triticum aestivum*). Following plantlet formation, determine the percentage of haploids among the newly formed plants by using acetocarmine squash preparations of the root tips.

2. With the aid of a hemocytometer, determine the number of pollen grains within a single anther taken from your experimental plant. What percentage of pollen grains formed embryoids from the anthers you cultured?

3. Repeat Experiment 1 with the agar medium supplemented with activated charcoal (2% w/v). Do you find a greater percent-

age of anthers forming embryoids compared to the original results? Is the total number of plantlets produced greater or smaller in comparison to the medium lacking the charcoal?

SELECTED REFERENCES

Anagnostakis, S. L. (1974). Haploid plants from anthers of tobacco: Enhancement with charcoal. *Planta 115*, 281–3.

Bajaj, Y. P. S. (1983). In vitro production of haploids. In *Handbook of plant cell culture*, vol. 1, *Techniques for propagation and breeding*, ed. D. A. Evans, W. R. Sharp, P. V. Ammirato, & Y. Yamada, pp. 228–87. New York: Macmillan.

Bajaj, Y. P. S., Reinert, J., & Heberle, E. (1977). Factors enhancing in vitro production of haploid plants in anthers and isolated microspores. In *La culture des tissus et des cellules des végétaux: Résultats généraux et réalisations practiques*, ed. R. J. Gautheret, pp. 47–58. Paris: Masson.

Bhojwani, S. S., & Bhatnagar, S. P. (1974). *The embryology of angiosperms.* Delhi: Vikas.

Burk, L. G., Gwynn, G. R., & Chaplin, J. F. (1972). Diploidized haploids: From aseptically cultured anthers of Nicotiana tabacum. *J. Hered. 63*, 355–60.

Collins, G. B. (1979). Cytogenetic techniques. In *Nicotiana: Procedures for experimental use*, ed. R. D. Durbin, pp. 17–22. Washington, D.C.: USDA Tech. Bull. No. 1586.

Darlington, D. C., & LaCour, L. F. (1976). *The handling of chromosomes*, 6th Ed. London: Allen & Unwin.

Devreux, M. (1970). New possibilities for the in vitro cultivation of plant cells. *EuroSpectra 9*, 105–10.

Devreux, M., & Laneri, V. (1974). Anther culture, haploid plant, isogenic line and breeding researches in Nicotiana tabacum L. In *Polyploidy and induced mutations in plant breeding*, pp. 101–7. Proceedings of the Eucarpia/FAO/IAEA Conference, Rome, Italy, 1972.

Dodds, J. H., & Reynolds, T. L. (1980). A scanning electron microscope study of pollen embryogenesis in Hyoscyamus niger. *Z. Pflanzenphysiol. 97*, 271–6.

Dunwell, J. M. (1976). A comparative study of environmental and developmental factors which influence embryo induction and growth in cultured anthers of Nicotiana tabacum. *Environ. Exp. Bot. 16*, 109–18.

Guha, S., & Maheshwari, S. C. (1966). Cell division and differentiation of embryos in the pollen grains of Datura in vitro. *Nature 212*, 97–8.

(1967). Development of embryoids from pollen grains of Datura in vitro. *Phytomorphology 17*, 454–61.

Horner, M., & Street, H. E. (1978). Problems encountered in the culture of isolated pollen of a burley cultivar of Nicotiana tabacum. *J Exp. Bot. 29*, 217–26.

Hu Han, H. T. Y., Tseng, C. C., Ouyang, T. W., & Ching, C. K. (1978). Application of anther culture to crop plants. In *Frontiers of plant tissue culture 1978*, ed. T. A. Thorpe, pp. 123-30. Calgary: IAPTC.

Jensen, C. J. (1974). Chromosome doubling techniques in haploids. In *Haploids in higher plants – Advances and potential*, ed. K. J. Kasha, pp. 153-90. Guelph: University of Guelph Press.

Kasperbauer, M. J., & Wilson, H. M. (1979). Haploid plant production and use. In *Nicotiana: Procedures for experimental use*, ed. R. D. Durbin, pp. 33-9. Washington, D.C.: USDA Tech. Bull. No. 1586.

Keller, W. A., & Stringham, G. R. (1978). Production and utilization of microspore derived haploid plants. In *Frontiers of plant tissue culture 1978*, ed. T. A. Thorpe, pp. 113-22. Calgary: IAPTC.

Kohlenbach, H. W., & Wernicke, W. (1978). Investigations on the inhibitory effect of agar and the function of active carbon in anther culture. *Z. Pflanzenphysiol. 86*, 463-72.

Martineau, B., Hanson, M. R., & Ausubel, F. M. (1981). Effect of charcoal and hormones on anther culture of *Petunia* and *Nicotiana*. *Z. Pflanzenphysiol. 102*, 109-16.

Nagmani, R., & Raghavan, V. (1983). Induction of embryogenic divisions in isolated pollen grains of *Hyoscyamus niger* in a single step method. *Z. Pflanzenphysiol 109*, 87-90.

Nitsch, J. P. (1969). Experimental androgenesis in *Nicotiana*. *Phytomorphology 19*, 390-404.

Raghavan, V. (1975). Induction of haploid plants from anther cultures of henbane. *Z. Pflanzenphysiol. 76*, 89-92.

(1978). Origin and development of pollen embryoids and pollen calluses in cultured anther segments of *Hyoscyamus niger*. *Am. J. Bot. 65*, 984-1002.

Reinert, J., & Bajaj, Y. P. S. (1977). Anther culture: Haploid production and its significance. In *Applied and fundamental aspects of plant cell, tissue, and organ culture*, ed. J. Reinert and Y. P. S. Bajaj, pp. 251-67. Berlin: Springer-Verlag.

Sangwan-Norreel, B. S. (1977). Androgenic stimulating factors in the anther and isolated pollen grain cultures of *Datura innoxia* Mill. *J. Exp. Bot. 28*, 843-52.

Sopory, S. K., Jacobsen, E., & Wenzel, G. (1978). Production of monoploid embryoids and plantlets in cultured anthers of *Solanum tuberosum*. *Plant Sci. Lett. 12*, 47-54.

Sunderland, N. (1971). Anther culture: A progress report. *Sci. Prog. (Oxford) 59*, 527-49.

(1977). Comparative studies of anther and pollen culture. In *Plant cell and tissue culture*, ed. W. R. Sharp, P. O. Larson, & V. Raghavan, pp. 203-19. Columbus: Ohio State University Press.

(1979). Towards more effective anther culture. *Newsl. IAPTC 27*, 1012.

Sunderland, N., & Cocking, E. C. (1978). Plant tissue culture in China: Major change ahead? *Nature 274*, 643-4.

Sunderland, N., & Dunwell, J. M. (1977). Anther and pollen culture. In *Plant tissue and cell culture*, 2d Ed., ed. H. E. Street, pp. 223–65. Oxford: Blackwell Scientific Publications.

Sunderland, N., & Roberts, M. (1977). New approaches to pollen culture. *Nature 270*, 236–8.

Sunderland, N., & Wicks, F. M. (1971). Embryoid formation in pollen grains of *Nicotiana tabacum*. *J. Exp. Bot. 22*, 213–26.

Tanaka, M., & Nakata, K. (1969). Tobacco plants obtained by anther culture and the experiment to get diploid seed from haploids. *Jap. J. Genet. 44*, 47–54.

Thomas, E., & Davey, M. R. (1975). *From single cells to plants*. London: Wykeham.

Tulecke, W. (1953). A tissue derived from the pollen of *Ginkgo biloba*. *Science 117*, 599–600.

Vasil, I. K., & Nitsch, C. (1975). Experimental production of pollen haploids and their uses. *Z. Pflanzenphysiol. 76*, 191–212.

Weatherhead, M. A., Bordon, J., & Henshaw, G. G. (1978). Some effects of activated charcoal as an additive to plant tissue culture media. *Z. Pflanzenphysiol. 89*, 141–7.

Wenzel, G., Hoffmann, F., & Thomas, E. (1977). Increased induction and chromosome doubling of androgenetic haploid rye. *Theor. Appl. Genet. 51*, 81–6.

Wernicke, W., & Kohlenbach, H. W. (1976). Investigations on liquid culture medium as a means of anther culture in *Nicotiana*. *Z. Pflanzenphysiol. 79*, 189–98.

(1977). Versuche zur kultur isolierter mikrosporen von *Nicotiana* and *Hyoscyamus*. *Z. Pflanzenphysiol. 81*, 330–40.

Wilson, H. M., Mix, G., & Foroughi-Wehr, B. (1978). Early microspore divisions and subsequent formation of microspore calluses at high frequency in anthers of *Hordeum vulgare* L. *J. Exp. Bot. 29*, 227–38.

PART IV

*Experimental: Isolated cells*

# 12

## Transdifferentiation of parenchyma cells to tracheary elements

"Cell differentiation" refers to a morphological and biochemical specialization of a given cell, as well as the developmental process leading to this unique condition. Some mature plant cells have the potential to "transdifferentiate," that is, to become a different type of cell without undergoing cell division. When stimulated by hormonal signals to transdifferentiate, quiescent cells start genetic reprogramming that ultimately leads to the creation of the new cell type. "Determination" refers to the resumption of reprogramming by which a cell becomes restricted to a new pathway of specialization (Roberts, 1976). Direct transdifferentiation of parenchymatous cells into tracheary elements provides the investigator with an excellent model system for the study of cytodifferentiation (Sugiyama & Komamine, 1990).

There are several advantages in using the xylem cell for studying cytodifferentiation. With the exception of sieve elements, few other cell types can be induced to form in vitro under experimentally controlled conditions. Xylem cells are easily identified with the light microscope because of the striking patterns of secondary-wall thickenings (Fig. 12.1). Tracheary element formation occurs at a relatively rapid and predictable rate, and a reasonable level of synchronous differentiation occurs in single-cell cultures of *Zinnia elegans* (Fukuda & Komamine, 1980a). In these mesophyll cell cultures xylogenesis can occur directly without cell division (Kohlenbach & Schmidt, 1975; Fukuda & Komamine, 1980b; Kohlenbach & Schöpke, 1981). Since xylem cell formation is terminal, resulting in cell autolysis, the experimental results cannot be confused with any additional differentiation process (Fukuda & Komamine, 1985). Also several biochemical markers for xylogenesis have been identified (Bolwell, 1985). The cultural requirements for the induction of xylem differentiation are relatively simple. Although a basal medium is usually employed, the minimal exogenous requirements for xylogenesis in most parenchyma explants are an auxin, a cytokinin, and a carbohydrate. Ethylene may also be a hormonal requirement (Miller & Roberts, 1984; Koritsas, 1988). Gibberellic acid interacts with other growth regulators during the in vitro differentiation of secondary xylem fibers (Aloni, 1982). A role for brassinosteroids has been suggested (Iwasaki & Shibaoka, 1991).

Let us briefly review the terminology of the water-conducting cells of the xylem tissue. "Tracheary elements," which are subdivided into "tracheids" and "vessel elements," transport water and

Figure 12.1. Contiguous tracheary elements usually display a remarkable similarity in the patterns of secondary-wall thickenings, which suggests that pattern determination during cytodifferentiation can be passed from cell to cell. The two tracheary elements in the photomicrograph differentiated in an explant of lettuce pith parenchyma cultured on a xylogenic medium. x475. (Roberts, 1976.)

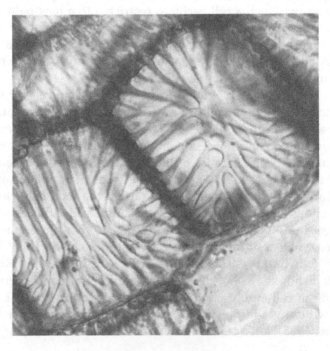

dissolved substances upward in the xylem. Both cell types are dead at maturity and possess pitted lignified cell walls. These two types differ in that tracheids have imperforate end walls, whereas the end walls of vessel elements have various types of perforations. A longitudinal column of vessel elements, interconnected end to end by perforation plates, is known as a "vessel" (Esau, 1977).

Cell types found in tissue cultures include vessel elements or wound vessel members (Roberts & Fosket, 1962), tracheids, and cells that resemble xylem fibers (Roberts, 1988). Xylem cells formed in vitro exhibit the following sequential stages of development. First, the cell acquires a competence to initiate genetic reprogramming, that is, it passes a point of determination. The cell may exhibit some degree of cell enlargement, followed by secondary wall deposition, lignification, autolysis of the cell contents, and selective dissolution of the cell wall (Roberts, 1976).

Since the student experiment will involve transdifferentiation in *Zinnia* mesophyll cells, let us review some information on this interesting system. Kohlenbach and Schmidt (1975) discovered that mesophyll cells isolated from the leaves of *Zinnia elegans* could be induced to transdifferentiate directly into tracheary elements after three days of culture in a medium containing 2,4-D and kinetin. Improvements in the experimental technique led to a high frequency and synchrony of xylem cell formation with 30–40% of the mesophyll cells differentiating within 60–80 hr of culture (Fukuda & Komamine, 1980a,b). During the past decade several aspects of transdifferentiation of cultured *Zinnia* cells have been examined by numerous research groups.

A majority of the tracheary elements were formed directly by cells in the first $G_1$ phase of the cell cycle, thus indicating that DNA replication in the S phase was not a prior requirement for transdifferentiation (Fukuda & Komamine, 1981a). The hypothesis was given that some minor amount of DNA synthesis, separate from S-phase replication, may be a requirement for transdifferentiation (Fukuda & Komamine, 1981b). A repair-type synthesis of nuclear DNA, driven by a combination of $\alpha$- and $\beta$-type polymerases, occurs in the differentiating cells. The inhibition of transdifferentiation by both 5-fluorodeoxyuridine and aphidicolin supports the hypothesis that repair-type DNA synthesis is a necessity for xylogenesis in this system (Sugiyama & Komamine, 1987a;

Sugiyama, Fukuda, & Komamine, 1990). Furthermore, inhibitors of ADP-ribosyltransferase blocked transdifferentiation, and this enzyme plays a role in DNA excision repair (Sugiyama & Komamine, 1987b).

Organization of the microtubules changes at an early stage of transdifferentiation. There is a biosynthesis of tubulin, a major component of microtubules (Fukuda, 1987, 1989a), and the number of microtubules increases rapidly within 24–48 hr of culture (Fukuda & Kobayashi, 1989). Immunofluorescence and ultrastructural studies have shown that microtubule arrays and the secondary wall bands are closely related. Microtubules are grouped and localized under the ridges of secondary wall bands formed transversely to the long axis of the differentiating cells (Burgess & Linstead, 1984; Falconer & Seagull, 1985a,b, 1986; Fukuda & Kobayashi, 1989). Actin filaments appear to play a role interacting with the microtubules in the orientation of the secondary wall (Kobayashi, Fukuda, & Shibaoka, 1988; Fukuda & Kobayashi, 1989). During secondary wall formation there is an increase in the levels of xylan and cellulose in the transdifferentiating cells (Ingold, Sugiyama, & Komamine, 1988). The presence of xylan in the secondary walls of the *Zinnia* cells was confirmed by an immunocytochemical technique (Northcote, Davey, & Lay, 1989). An ultrastructural study of transdifferentiating *Zinnia* cells revealed rosettes composed of six subunits in the fracture faces of Golgi cisternae. These rosettes were located over the secondary-wall thickenings and are thought to be complexes of cellulose-synthesizing enzymes that participate in wall formation (Haigler & Brown, 1986).

Several studies have been made on the lignification process occurring during transdifferentiation. Enzymes associated with lignin synthesis are considered markers of xylogenesis in *Zinnia* (Fukuda & Komamine, 1982). Coumarate–coenzyme A ligase activity was found to be a characteristic of transdifferentiating *Zinnia* cells (Church & Galston, 1988b). Transdifferentiation, however, is evidently independent of lignification. The inhibitor L-α-aminooxy-β-phenylpropionic acid blocked lignification but not the differentiation of tracheary elements in cultured *Zinnia* cells (Ingold, Sugiyama, & Komamine, 1990). Finally, many degradative enzymes play roles during the autolytic termination of the process. In addition to a nuclease specific for single-stranded nucleic acids, three RNase species were found to be specific for differentiating *Zinnia* mesophyll cells (Thelen & Northcote, 1989).

Additional information on transdifferentiation of *Zinnia* mesophyll cells into tracheary elements can be found in the reviews compiled by Fukuda and Komamine (1985), Fukuda (1989b), Sugiyama and Komamine (1990), and Church (1993).

In the present experiment the student will attempt to induce the transdifferentiation of single cells isolated from the leaf mesophyll tissue of *Zinnia elegans* seedlings. The instructor can suggest numerous ancillary experiments involving changes in growth regulators, growth regulator concentrations, and altered cultural conditions. A slightly different approach involving transdifferentiation in *Zinnia* leaf disks is outlined in the Appendix.

LIST OF MATERIALS

*Sterilization mode:* C, chemical; O, oven; A, autoclave

- C  1.5 g of first or second leaves excised from seedlings of *Zinnia elegans* L.
- –  300 $cm^3$ aqueous solution (1.0% v/v) commercial bleach containing a final concentration of approximately 0.05% (v/v) NaOCl
- O  400-$cm^3$ beakers (four); foil wrapped
- O  mortar and pestle (smooth porcelain recommended)
- O  stainless steel forceps enclosed in test tube
- O  9-cm Petri dishes (five); foil wrapped (sterile surface for cutting leaves)
- O  50-$cm^3$ Erlenmeyer flasks (five)
- O  5-$cm^3$ pipette
- A  85-$\mu$m nylon screen with 50-$cm^3$ beaker (filtration unit)
- A  disposable plastic centrifuge tubes
- A  125-$cm^3$ Erlenmeyer flasks, each containing 100 $cm^3$ DDH$_2$O (rinse water for three beakers) (six)
- A  Fukuda and Komamine's modified culture medium (1 liter for class use)
- A  0.2 M mannitol osmoticum (50 $cm^3$)
- C  scalpel; Bard–Parker No. 7 surgical knife handle with a No. 10 blade is recommended
- –  ethanol (80% v/v) dip
- –  ethanol (70% v/v) in plastic squeeze bottle
- –  methanol lamp
- –  interval timer

- heavy-duty aluminum foil (one roll)
- Tween 80
- Vermiculite; pots; Hyponex
- controlled-environment facility (25 °C); fluorescent lighting
- centrifuge (table model)
- Fuchs–Rosenthal modified hemocytometer
- Evans blue dye reagent compound microscope (100× magnification)
- microscope slides; cover slips

*Note:* If a Waring-type blender is used for tissue maceration, autoclave the pertinent parts.

PROCEDURE

*Plant material.* Germinate seeds of *Zinnia elegans* L. in sterile moist Vermiculite and water daily. Water with a 1.2 g/l solution of Hyponex four days after sowing (Church & Galston, 1988a). Although the cultivars Canary Bird and Envy are recommended, other cultivars will give comparable results. Grow the seedlings in a controlled environment facility at 25 °C with 16–8-hr light–dark photoperiods. A fluorescent light intensity of about 20W/m$^2$ or 5,000–10,000 lux is recommended, and the relative humidity should be maintained below 60%.

*Isolation of single mesophyll cells.* Harvest approximately 1.5 g of first leaves from 12–14-day-old seedlings and rinse in distilled water containing 0.2% (v/v) Tween 80. Surface sterilize the leaves for 10 min in a 1.0% aqueous solution of commercial bleach (0.05% NaOCl) and rinse three times with sterile DDH$_2$O. In a sterile Petri dish cut the leaves into small pieces measuring about 5 × 5 mm and macerate them in 5 cm$^3$ of 0.2 M mannitol with a mortar and pestle. An alternative technique is to macerate the leaves in a Waring-type blender after autoclaving whatever parts of the blender will come into contact with the sterile leaves. Add approximately 4.0 g of leaves to the container in 70 cm$^3$ of mannitol solution and blend for 30 sec at 10,000 rpm (Fukuda, pers. commun.). Wash the filtered cells by repeated centrifugation (150 $g$ × 2 min) and resuspend in the mannitol osmoticum. The cells are finally suspended in Fukuda and Komamine's (1980a) culture medium as modified by Sugiyama, Fukuda, and Komamine (1986). Although these workers resuspended the cells in 150 cm$^3$ of medium, the final volume will be determined by the cell density.

*Initial cell density.* For a high frequency of transdifferentiation it is recommended that the initial cell population density be within the range $0.1$–$1.0 \times 10^5$ cells/cm$^3$. A similar requirement is involved with the culture of isolated protoplasts (Chapter 13). The cell suspension should be diluted with the culture medium until the cell density falls within this range. The concentration in a given sample can be determined by the use of a Fuchs–Rosenthal hemocytometer modified for a field depth of 0.2 mm. Hemocytometers designed for counting red blood cells normally have a field depth of 0.1 mm, which is too shallow for large plant cells. See Chapter 17 for the calculations involved in determining cell counts.

*Culture protocol.* Culture the single-cell suspensions in the dark (25 °C) in 4-cm$^3$ aliquots in 50-cm$^3$ Erlenmeyer flasks on an orbital shaker (75 rpm), or 3.0-cm$^3$ aliquots in tubes (18 × 180 mm) on a rotating drum (10 rpm).

## RESULTS

A sharp increase in tracheary element number should occur between 48 and 72 hr of culture, and the number increases again from day 6 of culture (Fukuda & Komamine, 1980a). The highest proportion of tracheary elements, which is 30–40% of the total number of cells, is found three days after culture. The percent differentiation can be determined with the aid of Evans blue dye: Cells with an intact plasmalemma will exclude this biological stain; thus impermeability of the cell to the dye indicates a living cell.

$$\text{Percent differentiation} = \frac{\text{number of tracheary elements}}{\text{total cell number}} \times 100$$

Total cell number represents tracheary elements plus the living cells in the culture.

## QUESTIONS FOR DISCUSSION

1. What are some advantages of using isolated *Zinnia* mesophyll cells for the study of tracheary element differentiation?
2. Define the term "transdifferentiation."
3. What evidence suggests that some DNA synthesis may be a requirement prior to tracheary element differentiation?
4. Of what importance are microtubules to the formation of tracheary elements?
5. Discuss the growth regulator requirements for the in vitro induction of tracheary element differentiation.

## APPENDIX

*Induction of xylogenesis in pith parenchyma explants of lettuce (Lactuca sativa L.).* Remove the leaves from a head of Romaine (cos) lettuce, trim the lateral sides of the pith core until smooth, and slice off the basal end. Cut off the apex about 2 cm from the tip and rinse the core in tap water. Surface sterilize the core with hypochlorite followed by three rinses in $DDH_2O$ in the usual manner (Chapter 5, Procedure). Prepare cylindrical borings with a sterile cork borer (4 mm I.D.). Slice the borings transversely into segments approximately 2–3 mm in thickness. Transfer the disk-shaped explants to the surface of a xylogenic medium in culture tubes or Petri dishes. Culture the explants in a dark incubator at 25 °C. The following xylogenic medium will produce excellent results: MS basal salts, MS vitamin supplement, *myo*-inositol (100 mg/l), IAA (10 mg/l), kinetin (0.1 mg/l), sucrose (2.0% w/v), and Gelrite (0.2% w/v). Although the minimum time for xylogenesis in this system is about 4 days, plan to terminate the experiment after 7–10 days. At that time place the explants singly in small glass vials. Add 2 cm³ aqueous solution of NaOH (4% w/v), stopper the vials, and place them in a 50 °C oven overnight. This treatment clears and softens the tissue. Carefully remove the NaOH solution with a syringe, add about 2 cm aqueous solution of safranin O (0.04% w/v), and return the vials to the 50 °C oven. After 30 min withdraw the dye solution, fill the vials with 1 $N$ HCl, and return them to the oven for 1 hr. The acid destains the parenchyma cells, but the lignified tracheary elements retain the red dye. Remove the discolored HCl and add a fresh change of HCl for an additional hour. Remove the HCl and cover the explants by adding glycerol, in which they may remain immersed indefinitely. With the aid of a pair of glass dissecting needles, pick out a small fragment of tissue containing clumps or strands of tracheary elements for microscopic examination. Transfer the tissue fragment to a slide, add a cover glass, and squash the preparation with the tip of a pencil. Examine with a compound microscope at 100× magnification. For additional information see the original publication by Dalessandro and Roberts (1971).

*Induction of tracheary element differentiation in leaf disks of Zinnia.* Xylogenesis can be induced readily in cultured leaf disks removed from *Zinnia* seedlings (Church & Galston, 1989). Surface sterilize the first leaves of 10-day-old *Zinnia* seedlings. Remove leaf disks (4 mm in diameter) with the aid of a sterile paper punch (hand model). Rinse the leaf disks with several changes of hormone-free medium (Fukuda & Komamine, 1980a). Since cells differentiate only near the cut edges of unpeeled leaf disks, the lower epidermis should be peeled off with a pair of fine curved forceps. Culture the disks singly in sterile, multiwell tissue culture plates. Each well contains 0.5 ml Fukuda and Komamine's medium supplemented with various concentrations of $\alpha$-NAA and BAP. The disks are cultured on an orbital shaker (75 rpm) in the dark at 27°C. For visualization of the differentiated tracheary elements the disks can be cleared and stained with safranin O as described previously for the lettuce pith explants. Extensive transdifferentiation of mesophyll-derived tracheary elements, as well as differentiation of cells adjacent to the preexisting vascular tissues, occurred in the presence of $\alpha$-NAA (0.5 mM) plus BAP (0.5 mM). Increasing the concentrations of both growth regulators to 5.0 mM greatly enhanced the transdifferentiation of epidermal cells to tracheary elements (Church & Galston, 1989).

SELECTED REFERENCES

Aloni, R. (1982). Role of cytokinin in differentiation of secondary xylem fibers. *Plant Physiol. 70,* 1631-3.
Bolwell, G. P. (1985). Use of tissue cultures for studies on vascular differentiation. In *Plant cell culture: A practical approach,* ed. R. A. Dixon, pp. 107-26. Oxford: IRL Press.
Burgess, J., & Linstead, P. (1984). In vitro tracheary element formation:

Structural studies and the effect of triiodobenzoic acid. *Planta* 160, 481-9.

Church, D. L. (1993). Tracheary element differentiation in *Zinnia* mesophyll cell cultures. *Plant Growth Regulation* 12, 179-88.

Church, D. L., & Galston, A. W. (1988a). Kinetics of determination in the differentiation of isolated mesophyll cells of *Zinnia elegans* to tracheary elements. *Plant Physiol.* 88, 92-6.

  (1988b). 4-Coumarate : coenzyme A ligase and isoperoxidase expression in *Zinnia* mesophyll cells induced to differentiate into tracheary elements. *Plant Physiol.* 88, 679-84.

  (1989). Hormonal induction of vascular differentiation in cultured *Zinnia* leaf discs. *Plant Cell Physiol.* 30, 73-8.

Dalessandro, G., & Roberts, L. W. (1971). Induction of xylogenesis in pith parenchyma explants of *Lactuca*. *Am. J. Bot.* 58, 378-85.

Esau, K. (1977). *Anatomy of seed plants*, 2d Ed. New York: Wiley.

Falconer, M. M., & Seagull, R. W. (1985a). Immunofluorescent and Calcofluor White staining of developing tracheary elements in *Zinnia elegans* L. suspension cultures. *Protoplasma* 125, 190-8.

  (1985b). Xylogenesis in tissue culture: Taxol effects on microtubule reorientation and lateral association in differentiating cells. *Protoplasma* 128, 157-66.

  (1986). Xylogenesis in tissue culture: II. Microtubules, cell shape and secondary wall patterns. *Protoplasma* 133, 140-8.

Fukuda, H. (1987). A change in tubulin synthesis in the process of tracheary element differentiation and cell division of isolated *Zinnia* mesophyll cells. *Plant Cell Physiol.* 28, 517-28.

  (1989a). Regulation of tubulin degradation in isolated *Zinnia* mesophyll cells in culture. *Plant Cell Physiol.* 30, 243-52.

  (1989b). Cytodifferentiation in isolated single cells (invited article). *Bot. Mag. Tokyo* 102, 491-501.

Fukuda, H., & Kobayashi, H. (1989). Dynamic organization of the cytoskeleton during tracheary-element differentiation. *Dev. Growth Differ.* 31, 9-16.

Fukuda, H., & Komamine, A. (1980a). Establishment of an experimental system for the study of tracheary element differentiation from single cells isolated from the mesophyll of *Zinnia elegans*. *Plant Physiol.* 65, 57-60.

  (1980b). Direct evidence for cytodifferentiation to tracheary elements without intervening mitosis in a culture of single cells isolated from the mesophyll of *Zinnia elegans*. *Plant Physiol.* 65, 61-4.

  (1981a). Relationship between tracheary element differentiation and the cell cycle in single cells isolated from the mesophyll of *Zinnia elegans*. *Physiol. Plant.* 52, 423-30.

  (1981b). Relationship between tracheary element differentiation and DNA synthesis in single cells isolated from the mesophyll of *Zinnia elegans*: Analysis by inhibitors of DNA synthesis. *Plant Cell Physiol.* 22, 41-9.

(1982). Lignin synthesis and its related enzymes as markers of tracheary element differentiation in single cells isolated from the mesophyll of Zinnia elegans. *Planta* 155, 423–30.

(1985). Cytodifferentiation. In *Cell culture and somatic cell genetics of plants*, vol. 2, ed. I. K. Vasil, pp. 149–212. Orlando, Fla.: Academic Press.

Haigler, C. H., & Brown, R. M., Jr. (1986). Transport of rosettes from the Golgi apparatus to the plasma membrane in isolated mesophyll cells of *Zinnia elegans* during differentiation to tracheary elements in suspension culture. *Protoplasma* 134, 111–20.

Ingold, E., Sugiyama, M., & Komamine, A. (1988). Secondary wall formation: Changes in cell wall constituents during the differentiation of isolated mesophyll cells of *Zinnia elegans* to tracheary elements. *Plant Cell Physiol.* 29, 295–303.

(1990). 1-$\alpha$-Aminooxy-$\beta$-phenylpropionic acid inhibits lignification but not the differentiation to tracheary elements of isolated mesophyll cells of *Zinnia elegans*. *Physiol. Plant.* 78, 67–74.

Iwasaki, T., & Shibaoka, H. (1991). Brassinosteriods act as regulators of tracheary-element differentiation in isolated *Zinnia* mesophyll cells. *Plant Cell Physiol.* 32, 1007–14.

Kobayashi, H., Fukuda, H., & Shibaoka, H. (1988). Interrelation between the spatial disposition of actin filaments and microtubules during the differentiation of tracheary elements in culture *Zinnia* cells. *Protoplasma* 143, 29–37.

Kohlenbach, H. W., & Schmidt, B. (1975). Cytodifferentiation in mode of direct transformation of isolated mesophyll cells to tracheids. *Z. Pflanzenphysiol.* 75, 369–91.

Kohlenbach, H. W., & Schöpke, C. (1981). Cytodifferentiation to tracheary elements from isolated mesophyll protoplasts of *Zinnia elegans*. *Naturwissenschaften* 68, 576–7.

Koritsas, V. M. (1988). Effect of ethylene and ethylene precursors on protein phosphorylation and xylogenesis in tuber explants of *Helianthus tuberosus* (L.). *J. Exp. Bot.* 39, 375–86.

Miller, A. R., & Roberts, L. W. (1984). Ethylene biosynthesis and xylogenesis in *Lactuca* pith explants cultured in vitro in the presence of auxin and cytokinin: The effect of ethylene precursors and inhibitors. *J. Exp. Bot.* 35, 691–8.

Northcote, D. H., Davey, R., & Lay, J. (1989). Use of antisera to localize callose, xylan and arabinogalactan in the cell-plate, primary and secondary walls of plant cells. *Planta* 178, 353–66.

Roberts, L. W. (1976). *Cytodifferentiation in plants: Xylogenesis as a model system*. Cambridge: Cambridge University Press.

(1988). Evidence from wound responses and tissue cultures. In *Vascular differentiation and plant growth regulators*, ed. L. W. Roberts, P. B. Gahan, & R. Aloni, pp. 63–88. Heidelberg: Springer–Verlag.

Roberts, L. W., & Fosket, D. E. (1962). Further experiments on wound vessel formation in stem wounds of *Coleus*. *Bot. Gaz.* 123, 247–54.

Sugiyama, M., Fukuda, H., & Komamine, A. (1986). Effects of nutrient limitation and γ-irradiation on tracheary element differentiation and cell division in single mesophyll cells of *Zinnia elegans*. *Plant Cell Physiol.* 27, 601–6.

(1990). Characteristics of the inhibitory effect of 5-fluorodeoxyuridine on cytodifferentiation into tracheary elements of isolated mesophyll cells of *Zinnia elegans*. *Plant Cell Physiol.* 31, 61–7.

Sugiyama, M., & Komamine, A. (1987a). Relationship between DNA synthesis and cytodifferentiation to tracheary elements. *Oxford Surv. Plant Mol. Cell Biol.* 4, 343–6.

(1987b). Effects of inhibitors of ADP-ribosyltransferase on the differentiation of tracheary elements from isolated mesophyll cells of *Zinnia elegans*. *Plant Cell Physiol.* 28, 541–4.

(1990). Transdifferentiation of quiescent parenchymatous cells into tracheary elements. *Cell Differ. Dev.* 31, 77–87.

Thelen, M. P., & Northcote, D. H. (1989). Identification and purification of a nuclease from *Zinnia elegans* L.: A potential molecular marker for xylogenesis. *Planta* 179, 181–95.

# 13

## Isolation and culture of protoplasts

Isolated protoplasts have been described as "naked" plant cells because the cell wall has been experimentally removed by either a mechanical or an enzymatic process. The isolated protoplast is unusual because the outer plasma membrane is fully exposed and is the only barrier between the external environment and the interior of the living cell (Cocking, 1972; Evans & Cocking, 1977). Despite technical difficulties that have limited their potential use in some investigations, protoplasts are currently utilized in several areas of study (Galun, 1981):

1. Two or more protoplasts can be induced to fuse and the fused product carefully nurtured to produce a hybrid plant. Although this phenomenon has been observed repeatedly, fusion has not been achieved with the isolated protoplasts of some species. In some cases, hybrids that cannot be produced by conventional plant genetics because of sexual or physical incompatibility can be formed by somatic cell fusion (Power & Cocking, 1971; Cocking, 1977a; Dudits et al., 1979; Galun & Aviv, 1986). The regeneration of *Atropa belladonna* plants from single isolated protoplasts is shown in Figure 13.1.
2. After removal of the cell wall, the isolated protoplast is capable of ingesting "foreign" material into the cytoplasm by a process similar to endocytosis as described for certain animal cells and protozoans. Experiments are in progress on the introduction of nuclei, chloroplasts, mitochondria, DNA, plasmids, bacteria, viruses, and polystyrene beads into protoplasts (Davey & Cocking, 1972;

168     IV  Experimental: Isolated cells

Carlson, 1973; Cocking, 1977b; Cress, 1982; Lurquin & Sheehy, 1982; Dodds & Bengochea, 1983).
3. The cultured protoplast rapidly regenerates a new cell wall and this developmental process offers a novel system for the study of wall biosynthesis and deposition (Willison & Cocking, 1972; Grout, Willison, & Cocking, 1973).
4. Populations of protoplasts can be studied as a single cellular system; that is, their manipulation is similar to that of microorganisms (Evans & Cocking, 1977). Microbio-

Figure 13.1. Sequence of development of a plantlet of *Atropa belladonna* from a single isolated protoplast. (a) A single isolated protoplast. (b) Cell wall regeneration and initiation of cell division. (c,d) Development of cell aggregates. (e) Appearance of embryoids on surface of callus. (f) Formation of plantlet on agar medium. (Courtesy of H. Lörz.)

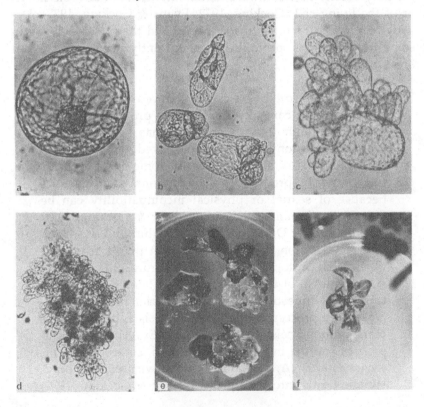

logical methods have been developed for the selection of mutant cell lines and the cloning of cell populations.

Most of the early research was conducted with the spongy and palisade mesophyll tissue obtained from mature leaves of *Nicotiana* and *Petunia*. With the improvement of techniques it has become possible to isolate protoplasts from a wide range of plant genera and tissues; for example, anthers of *Pelargonium* (Abo El-Nil & Hildebrandt, 1971), $C_3$ and $C_4$ plants (Kanai & Edwards, 1973), *Solanum tuberosum* (Upadhya, 1975), *Citrus* (Vardi, Spiegel-Roy, & Galun, 1975), callus of *Gosspium hirsutum* (Bhojwani, Cocking, & Power, 1977), and Crassulacean-acid-metabolism (CAM) plants (Dodds, 1980).

One must remember that a chief function of the cell wall is to exert a wall pressure on the enclosed protoplast and thus prevent excessive water uptake leading to bursting of the cell. Before the cell wall is removed the cell must be bathed in an isotonic plasmolyticum, which is carefully regulated in relation to the osmotic potential of the cell. In general, mannitol or sorbitol (13% w/v) has given satisfactory results. A technique involving the use of a relatively low osmotic potential with sucrose (0.2 $M$) and polyvinylpyrrolidone (2% w/v) has been reported (Shepard & Totten, 1975). In developing an original technique it may be advantageous to test a range of mannitol concentrations of 8–15% (w/v) (Evans & Cocking, 1977). The latter authors have pointed out that preparations bathed in a plasmolyticum of too low a concentration may lead to multinucleate protoplasts owing to the spontaneous fusion of two or more protoplasts during the isolation procedure.

As mentioned previously, protoplasts can be released from the cell wall by either a mechanical or an enzymatic process. The mechanical approach involves cutting a plasmolyzed tissue in which the protoplasts have shrunk and pulled away from the cell wall. Subsequent deplasmolysis results in expansion and release of the protoplasts from the cut ends of the cells. In practice this technique is difficult, and the yield of viable protoplasts is meager. One advantage, however, is that the complex and often deleterious effects of the wall-degrading enzymes on the metabolism of the protoplasts are eliminated. Nearly all the protoplast-isolation work since the early 1960s has been performed with enzymatic procedures. By using enzymes one obtains a high yield ($2\text{--}5 \times 10^6$ protoplasts/g leaf tissue) of uniform protoplasts after removal of cellular debris (Evans & Cocking, 1977). The basic technique consists of the following:

1. surface sterilization of leaf samples;
2. rinsing in a suitable osmoticum;
3. peeling off the lower epidermis or slicing the tissue to facilitate enzyme penetration;
4. sequential or mixed-enzyme treatment;
5. purification of the isolated protoplasts by removal of enzymes and cellular debris; and, finally,
6. transfer of the protoplasts to a suitable medium with the appropriate cultural conditions (see Figs. 13.2 and 13.3).

The plant cell wall consists of a complex mixture of cellulose, hemicellulose, pectin, and lesser amounts of protein and lipid. Because of the chemical bonding of these diverse constituents, a mixture of enzymes would appear necessary to degrade the system effectively. Cellulose is a polymer consisting of subunits of D-glucose. Xylans form the bulk of the hemicellulose fraction in angiosperms (Northcote, 1972). These polymers consist, however, of several monosaccharides in addition to xylose.

Pectins are polysaccharides containing the sugars galactose, arabinose, and the galactose derivative galacturonic acid (Northcote, 1974). Protoplast isolation is achieved by using cellulase in combination with pectinase and hemicellulase. Most commercial preparations of these enzymes are isolated from microorganisms and often exhibit a variety of enzymatic activities. They may contain ribonucleases, proteases, and several other toxic enzymes. Because of the presence of these deleterious enzymes, as well as other harmful substances, several purification procedures have been developed. Although these techniques are beyond the scope of the present experiment, interested students can find detailed instructions in Patnaik, Wilson, and Cocking (1981), Constabel (1982), and Evans and Bravo (1983). There have been two approaches to the use of wall-degrading enzymes. In the "mixed-enzyme method" both pectinase and cellulase–hemicellulase are applied simultaneously, whereas the "sequential method" involves treatment of the leaf material with pectinase to loosen the cells, followed by a cellulase–hemicellulase digestion. Although both methods have certain advantages and disadvantages (Bajaj, 1977; Evans & Cocking, 1977), the mixed-enzyme technique will be used in the present experiment.

Each cellular system chosen for protoplast isolation presents its own unique problems in terms of isolation and culture procedures. Additional information on the selection of wall-degrading enzymes, as well as other variations in procedure, can be found in the Appendix. In this chapter the standard procedure for obtaining isolated protoplasts from mature leaf mesophyll tissue is described. This technique can be employed with a reasonable degree of success on other tissues. In the following experiment the student will isolate, purify, and culture mesophyll protoplasts. An attempt will be made to regenerate plantlets from the protoplast-seeded callus.

LIST OF MATERIALS

*Sterilization mode:* C, chemical; O, oven; A, autoclave; U, ultrafiltration

- C mature leaves from suitable plant (Appendix)
- – paring knife
- – 250 $cm^3$ aqueous solution (2.5% v/v) commercial bleach containing 2 $cm^3$ of detergent or wetting agent (1% v/v)
- C scalpel
- O forceps
- U 10 $cm^3$ enzyme solution consisting of Macerozyme R-10 (0.5% w/v) plus cellulase Onozuka R-10 (2.0% w/v) dissolved in mannitol (13% w/v), pH 5.4
- A 45-$\mu$m-pore-size nylon mesh
- A Millipore filter holder or similar ultrafilter equipped with a membrane filter, 0.45-$\mu$m pore size
- A 125-$cm^3$ Erlenmeyer flasks, each containing 100 $cm^3$ $DDH_2O$ (six)
- A 250-$cm^3$ Erlenmeyer flask containing 200 $cm^3$ MS medium plus mannitol (13% w/v)
- A 250-$cm^3$ Erlenmeyer flask containing 200 $cm^3$ MS–mannitol (13% w/v) plus agar (1.5% w/v), 40 °C
- O 2-$cm^3$ pipettes (six)
- O 10-$cm^3$ pipettes with wide bore (six)
- O Pasteur pipettes with wide bore (six)

O  centrifuge tubes capped with aluminum foil, 6 × 16 mm
O  white tile
−  Fuchs–Rosenthal hemocytometer, 0.2-mm field depth
−  ethanol (80% v/v) dip
−  ethanol (70% v/v) in plastic squeeze bottle
−  methanol lamp
−  bench-top centrifuge, swinging-bucket type
−  Evans blue dye reagent
−  fluorescein diacetate
−  fluorescence microscopy
−  Parafilm (one roll)
−  heavy-duty aluminum foil (one roll)
−  hypochlorite–detergent solution

PROCEDURE

1. Mature healthy leaves are removed from the plants and rinsed briefly in tap water. The interior of the transfer chamber is wiped with a tissue soaked in ethanol (70% v/v), and all subsequent procedures are conducted under aseptic conditions. The leaves will be treated as shown in Figure 13.2. Immerse the leaves in the hypochlorite–detergent solution for 10 min. As mentioned in the previous experiments, the UV lamp in the chamber should not be turned on during the hypochlorite sterilization. Rinse the leaves three times in order to remove all traces of the hypochlorite solution. Although the leaves may be rinsed in sterile $DDH_2O$, it is preferable to rinse them in a culture medium adjusted to the osmolality and pH of the enzyme solution to be used.

2. The waxy cuticle covering the leaves restricts access of the enzyme solution to the mesophyll cells. While the leaves are in the final rinse, the lower epidermis is peeled from the leaves with pointed forceps (as an alternative, leaves may be pricked). The leaves are easier to peel if held on a sterile white tile. The point of the forceps is inserted at the junction of the main vein or midrib, and the epidermal layer is stripped toward the edge of the lamina. If this approach is unsuccessful, the leaves should be allowed to wilt before another attempt is made. If further attempts to peel the leaf fail, then the lower epidermis is scored several times with a scalpel to aid the penetration of the enzymes. Cut the leaves into small sections and transfer approximately 1 g of peeled leaf

strips to a Petri dish (100 × 15 mm) containing 10 cm$^3$ of enzyme solution that has been sterilized by membrane filtration (0.45 μm). Although the types and concentrations of enzymes that may be employed are discussed later, it is suggested for the initial experiment that the student use Macerozyme R-10 (0.5% w/v) plus cellulase Onozuka R-10 (2.0% w/v) dissolved in mannitol (13% w/v) at pH 5.4 (Power & Cocking, 1971). Seal the Petri dishes with Parafilm and wrap them with aluminum foil. Usually the leaf material is incubated in the enzyme solution overnight (12–18 hr, 25 °C), although the mesophyll cells should be in contact with the enzymes for as short a time as possible. The leaf strips are then teased gently with forceps to release the protoplasts.

3. The protoplasts are purified by a combination of filtration, centrifugation, and washing (Fig. 13.3). First, the enzyme solution

Figure 13.2. Basic technique for the isolation of protoplasts from an excised leaf. The leaf is surface sterilized, rinsed repeatedly in sterile distilled water, and the cells are plasmolyzed in a solution of mannitol. The lower epidermal layer is stripped from the leaf to enhance enzyme penetration into the mesophyll tissue. Following treatment with one or more wall-degrading enzymes, a crude suspension of mesophyll protoplasts is obtained.

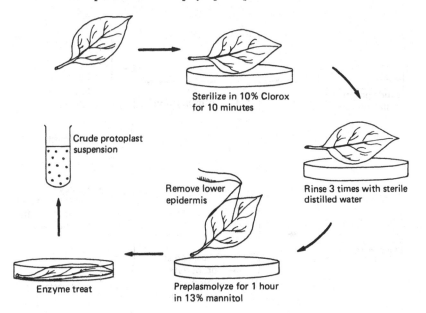

containing the protoplasts is filtered through a nylon mesh to remove undigested tissue, cell clumps, and cell wall debris. Transfer the filtrate to a centrifuge tube and spin it at 75 $g$ for 5 min. The debris in the supernatant is carefully removed with a Pasteur pipette, the protoplasts having formed a pellet at the base of the tube. The protoplasts are resuspended in 10 cm$^3$ culture medium (complete MS plus mannitol, 13% w/v), and the process is then repeated twice. The resuspension of the protoplasts must be carried out with considerable care with a wide-bore pipette (10 cm$^3$) in order to avoid injury. After the protoplasts have been examined for density and viability, they are ready for culture.

Figure 13.3. Purification procedure for isolated protoplasts. The crude protoplast suspension is filtered through a nylon mesh (45-µm pore size), and the filtrate is centrifuged for 5 min at 75 $g$. The supernatant, carefully removed by Pasteur pipette, is discarded. The protoplasts, resuspended in 10 cm$^3$ of fresh culture medium, are again centrifuged. Once again the supernatant is removed. The centrifugation–resuspension process is conducted three times. Before transfer of the protoplasts to a culture medium, the preparation is examined for protoplast density and viability (not shown).

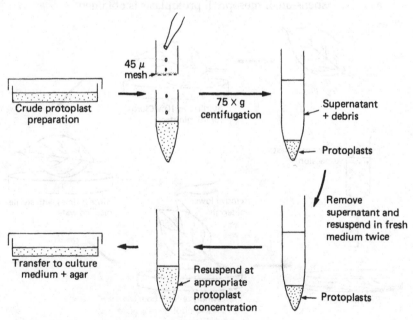

*Determination of protoplast density and viability.* Before the isolated protoplasts can be placed into culture it is necessary to examine them for viability with fluorescein diacetate. This dye, which accumulates only inside the plasmalemma of viable protoplasts, can be detected with fluorescence microscopy (Larkin, 1976; Strange, Pippard, & Strobel, 1982). Another staining method employs Evans blue. Intact viable protoplasts, by virtue of their intact plasmalemmas, are capable of excluding this biological stain: Impermeability of the cell to this dye presumably indicates a living cell (Kanai & Edwards, 1973). In addition, cyclosis or protoplasmic streaming has been reported to be a measure of viability (Raj & Herr, 1970).

Protoplasts have a maximum as well as a minimum plating density for growth. The optimum plating efficiency for tobacco protoplasts is about $5 \times 10^4$ protoplasts/cm$^3$; the protoplasts fail to divide when plated at one-tenth of this concentration (Evans & Cocking, 1977). Other protoplasts have similar ranges; for example, isolated protoplasts of *Petunia* have an optimum plating density of $2.5 \times 10^4$ (Power et al., 1976). Using a Fuchs–Rosenthal hemocytometer with a modified field depth of 0.2 mm (Chapter 12), it is possible to determine the concentration of protoplasts in a given sample and adjust it to the appropriate level. In the present experiment we shall assume a value similar to that for tobacco protoplasts. Because the protoplast preparation will be diluted by an equal quantity of agar-containing medium, the sample should be adjusted to a concentration of $10^5$ protoplasts/cm$^3$.

*Culture of protoplasts.* Protoplasts have been cultured in several ways; for example, in hanging-drop cultures (Gleba, 1978), in microculture chambers (Vasil & Vasil, 1973), and in a soft agar (0.75% w/v) matrix (Nagata & Takebe, 1971). The agar-embedding technique is one of the better methods as it ensures support for the protoplasts and permits observation of their development. The suspension of isolated protoplasts is adjusted with the hemocytometer by the addition of the culture medium plus mannitol (13% w/v) to yield a concentration of $10^5$ protoplasts/cm$^3$. Five cubic centimeters of this liquid suspension of protoplasts is added to a Petri dish (100 × 15 mm). To this dish is then added 5 cm$^3$ of the complete MS medium plus mannitol (13%, w/v); this latter solution also contains warm agar (1.5%, w/v) in the sol state at

40 °C. Care must be taken that the temperature does not exceed 45 °C (Bajaj, 1977). The two aliquots are mixed by swirling the dish, and the final result is 10 cm$^3$ of a medium containing 5 × 10$^4$ protoplasts/cm$^3$. The culture of embedded protoplasts is incubated at 25 °C in the presence of a dim white light.

*Regeneration of plants from the protoplasts.* Once the protoplasts have regenerated a cell wall, they undergo cell division and form a callus. This callus can be subcultured to plates or flasks containing a freshly prepared medium. If the callus of some species is transferred to a medium lacking both mannitol and auxin, embryogenesis begins on the callus after about three to four weeks (Kameya & Uchimiya, 1972; Lörz, Potrykus, & Thomas, 1977; Lörz & Potrykus, 1979). These embryoids, dissected from the callus, are nurtured in the same manner as those produced by somatic embryogenesis (Chapter 8) or by anther culture (Chapter 11). With proper care and attention the embryoids will develop into seedlings and eventually grow into mature plants.

*Preparation of the culture medium.* Follow the procedure outlined in Chapter 4 for the preparation of 1 liter of MS medium described for callus initiation. Because this medium contains an auxin and a cytokinin, it will readily induce growth and development in the isolated protoplasts. Add a suitable quantity of mannitol to the MS medium to yield a final concentration of 13% (w/v) mannitol. The function of mannitol is to prevent the osmotic lysis of the isolated protoplasts. Some of the medium also requires the addition of agar (1.5% w/v) before sterilization.

RESULTS

If green mesophyll tissue has been used for the isolation of protoplasts, the individual protoplasts will be readily visible with the use of bright field optics. This will also be true for pigmented protoplasts, such as those from petals or pigmented storage tubers (*Beta vulgaris, Daucus carota*). If colorless tissue has been used, the naked protoplasts will be visible with the aid of phase-contrast optics.

When first placed into culture, the isolated protoplasts are spherical because of the lack of a rigid cell wall (see Fig. 13.1a).

Once they are in culture on a suitable medium, a cell wall is quickly re-formed. After five to seven days some of the cells begin to undergo cell division (Fig. 13.1b). Repeated cell division gives rise to clumps of cells (Fig. 13.1c,d), which eventually produce callus masses visible to the naked eye. Once callus is sufficiently large to be manipulated, it may be subcultured to a medium lacking mannitol and auxin. The latter medium induces the formation of embryoids (Fig. 13.1e), which may be nurtured to maturity (Fig. 13.1f).

### QUESTIONS FOR DISCUSSION

1. What would be the result of transferring the isolated protoplasts to distilled water?
2. What is the advantage of using mannitol in preference to sucrose as an osmoticum?
3. List some of the possible applications of isolated plant protoplasts to the field of agriculture.
4. What are some advantages of using a mechanical technique over enzymatic digestion in the isolation of protoplasts?
5. What kinds of "foreign" material have been introduced into isolated protoplasts?

### APPENDIX

*Selection of plant material.* With protoplast isolation it is important that considerable thought be given to the selection of an appropriate plant. In addition, the suitability of a given leaf for protoplast isolation will be influenced by its age, its position on the plant, and environmental conditions (Constabel, 1982). A reasonable degree of success can be achieved by employing leaves from either *Nicotiana tabacum* L. var. Xanthi (tobacco), *Petunia* spp., or *Hyoscyamus niger* (henbane). It is inadvisable, however, to employ any systemic fungicides or insecticides during the growth of these plants. Although protoplasts can be isolated from callus and cell suspension cultures, the techniques are not clearly defined. In beginning experiments involving protoplast isolation, it is advisable to select a leaf mesophyll tissue at least until the techniques are mastered.

*Selection of wall-degrading enzymes.* At the present time a wide range of enzyme preparations are available for the digestion of the cell wall. Depending on the plant source and cellular type, varying concentrations of cellulase, hemicellulase, and pectinase will be required. A list of commercial preparations indicating the major enzymatic activity of each is given in Table 13.1. Typically, an enzyme preparation is chosen from each category. Some examples of mixtures of enzyme preparations that have been formulated for different plant material are given in Table 13.2. Additional formulations for various plant tissues can be found in the review by Evans and Bravo (1983).

*Variability of cultural conditions.* Results may differ depending on the commercial agar employed and the concentration used in the culture medium. For example, an agar concentration as low as 0.2% (w/v) has given good results (Binding, 1974). Protoplast division is sensitive to temperature, and some mesophyll protoplasts

Table 13.1. *Some enzyme preparations exhibiting wall-degrading activity classified according to major function*

| Cellulases | Hemicellulases | Pectinases |
|---|---|---|
| cellulase Onozuka (R-10, RS) | Rhozyme HP-150 | Macerase |
| Cellulysin | hemicellulase (Sigma) | Macerozyme R-10 |
| Meicelase (CESB, CMB) | | Pectolyase Y-23 |
| Driselase | | |

*Note:* Commercial suppliers are given at the end of the book.

Table 13.2. *Examples of combinations of enzyme mixtures used successfully for the preparation of protoplasts*

| Plant material | Enzyme mixture | Reference |
|---|---|---|
| *Hemerocallis* (cell suspension) | Cellulysin (1.0%), Rhozyme (1.0%), Macerase (0.5%) | Fitter & Krikorian (1983) |
| *Pisum sativum* L. (leaf mesophyll) | Onozuka R-10 (2.0%), Driselase (2.0%), Rhozyme (2.0%), pectinase (1.0%) | Constabel (1982) |
| *Solanum* sp. (leaf mesophyll) | Onozuka R-10 (1.0%), Macerozyme R-10 (0.5%), Pectolyase Y-23 (0.013%) | O'Hara & Henshaw (1982) |
| *Medicago sativa* (root; cotyledon) | Meicelase (4.0%), Rhozyme (2.0%), Macerozyme R-10 (0.03%) | Lu et al. (1982) |

may respond better at incubation temperatures higher than 25 °C. Another cultural variable is light. Although the present experiment was conducted with dim light, the plating efficiency may improve with higher light intensities after the first 48 hr of culture. On the other hand, some mesophyll protoplasts respond better in complete darkness (Evans & Cocking, 1977). Another interesting possibility is enrichment of the culture medium with various organic growth factors (e.g., coconut water, carbohydrates, organic acids, and casamino acids) (Kao & Michayluk, 1975).

*Role of divalent cations in membrane stability.* During the initial period of isolation and culture, protoplasts appear to benefit from a higher level of divalent cations than is normally present in a typical tissue culture medium. The MS medium may be supplemented with $CaCl_2 \cdot 2H_2O$ (5–10 m$M$) alone or in some combination with $CaH_4(PO_4)_2 \cdot H_2O$ (Constabel, 1982). Mesophyll protoplasts isolated from *Petunia* grew better, according to Binding (1974), with the medium supplemented with a combination of $CaCl_2$ (5 m$M$) and $MgSO_4$ (4 m$M$).

*Observations on cell wall regeneration.* Mesophyll protoplasts start to regenerate a new cell wall within a few hours following isolation, although it may take several days to complete wall biosynthesis. These initial events occurring on the surface of the plasmalemma can be observed microscopically by using Calcofluor White M2R, purified (Polysciences Inc.). This white dye binds to wall material and exhibits fluorescence on irradiation with blue light. The regenerating cells are incubated in 0.1% (w/v) Calcofluor dissolved in the appropriate osmoticum for 5 min (Evans & Cocking, 1977). After rinsing to remove excess dye, the protoplasts can be examined microscopically. Cellulose layers will fluoresce when irradiated with UV light at 366 nm (Constabel, 1982). A mercury vapor lamp with excitation filter BG12 and suppression filter K510 may be employed as a light source.

## SELECTED REFERENCES

Abo El-Nil, M. M., & Hildebrandt, A. C. (1971). Geranium plant differentiation from callus. *Am. J. Bot. 58,* 475.

Bajaj, Y. P. S. (1977). Protoplast isolation, culture and somatic hybridization. In *Applied and fundamental aspects of plant cell, tissue, and organ culture*, ed. J. Reinert & Y. P. S. Bajaj, pp. 467–96. Berlin: Springer-Verlag.

Bhojwani, S. S., Cocking, E. C., & Power, J. B. (1977). Isolation, culture and division of cotton callus protoplasts. *Plant Sci. Lett. 8*, 85–9.

Binding, H. (1974). Regeneration von haploiden und diploiden Pflanzen aus Protoplasten von *Petunia hybrida*. *Z. Pflanzenphysiol. 74*, 327–56.

Carlson, P. S. (1973). The use of protoplasts in genetic research. *Proc. Natl. Acad. Sci. USA 70*, 598–602.

Cocking, E. C. (1972). Plant cell protoplasts, isolation and development. *Annu. Rev. Plant Physiol. 23*, 29–50.

(1977a). Protoplast fusion: Progress and prospects for agriculture. *Span 20*, 5–8.

(1977b). Uptake of foreign genetic material by plant protoplasts. *Int. Rev. Cytol. 48*, 323–41.

Constabel, F. (1982). Isolation and culture of plant protoplasts In *Plant tissue culture methods*, 2d rev. Ed., ed. L. W. Wetter & F. Constabel, pp. 38–48. Saskatoon: National Research Council of Canada, Prairie Regional Laboratory.

Cress, D. E. (1982). Uptake of plasmid DNA by protoplasts from synchronized soybean cell suspension cultures. *Z. Pflanzenphysiol. 105*, 467–70.

Davey, M. R., & Cocking, E. C. (1972). Uptake of bacteria by isolated higher plant protoplasts. *Nature 239*, 455–6.

Dodds, J. H. (1980). Isolation of protoplasts from the CAM plant *Umbilicus pendulinus*. *Z. Pflanzenphysiol. 96*, 177–9.

Dodds, J. H., & Bengochea, T. (1983). Principles of plant genetic engineering. *Outlook Agric. 12*, 16–21.

Dudits, D., Hadlaczky, G. Y., Bajszar, G. Y., Koncz, C. S., Lazar, G., & Horvath, G. (1979). Plant regeneration from intergeneric cell hybrids. *Plant Sci. Lett. 15*, 101–12.

Evans, D. A., & Bravo, J. E. (1983). Protoplast isolation and culture. In *Handbook of plant cell culture*, vol. 1, *Techniques for propagation and breeding*, ed. D. A. Evans, W. R. Sharp, P. V. Ammirato, & Y. Yamada, pp. 124–76. New York: Macmillan.

Evans, P. K., & Cocking, E. C. (1977). Isolated plant protoplasts. In *Plant tissue and cell culture*, ed. H. E. Street, pp. 103–35. Oxford: Blackwell Scientific Publications.

Fitter, M. S., & Krikorian, A. D. (1983). *Plant protoplasts: Some guidelines for their preparation and manipulation in culture.* LaJolla: Calbiochem-Behring.

Galun, E. (1981). Plant protoplasts as physiological tools. *Ann. Rev. Plant Physiol. 32*, 237–66.

Galun, E., & Aviv, D. (1986). Organelle transfer. *Methods Enzymol. 118*, 595–611.

Gleba, Y. Y. (1978). Microdroplet culture: Tobacco plants from single mesophyll protoplasts. *Naturwissenschaften 65*, 158–9.

Grout, B. W. W., Willison, J. H. M., & Cocking, E. C. (1973). Interactions at the surface of plant protoplasts, an electrophoretic and freeze-etch study. *Bioenergetics* 4, 311–28.
Kameya, T., & Uchimiya, H. (1972). Embryoids derived from isolated protoplasts of carrot. *Planta* 103, 356–60.
Kanai, R., & Edwards, G. E. (1973). Purification and enzymatically isolated mesophyll protoplasts from $C_3$, $C_4$ and CAM plants using an aqueous dextran–polyethylene glycol two phase system. *Plant Physiol.* 52, 484–90.
Kao, K. N., & Michayluk, M. R. (1975). Nutrient requirements for growth of *Vieia hajastana* cells and protoplasts at a very low population density in liquid media. *Planta* 126, 105–10.
Larkin, P. J. (1976). Purification and viability determinations of plant protoplasts. *Planta* 128, 213–16.
Lörz, H., & Potrykus, I. (1979). Regeneration of plants from mesophyll protoplasts of *Atropa belladonna*. *Experimentia* 35, 313–14.
Lörz, H., Potrykus, I., & Thomas, E. (1977). Somatic embryogenesis from tobacco protoplasts. *Naturwissenschaften* 64, 439–40.
Lu, D. Y., Davey, M. R., Pental, D., & Cocking, E. C. (1982). Forage legume protoplasts: Somatic embryogenesis from protoplasts of seedling cotyledons and roots of *Medicago sativa*. In *Plant tissue culture 1982*, ed. A. Fujiwara, pp. 597–8. Tokyo: Japanese Association for Plant Tissue Culture.
Lurquin, P. F., & Sheehy, R. E. (1982). Binding of large liposomes to plant protoplasts and delivery of encapsulated DNA. *Plant Sci. Lett.* 25, 133–46.
Nagata, T., & Takebe, I. (1971). Plating of isolated tobacco mesophyll protoplasts on agar medium. *Planta* 99, 12–20.
Northcote, D. H. (1972). Chemistry of the plant cell wall. *Annu. Rev. Plant Physiol.* 23, 113–32.
(1974). *Differentiation in higher plants*. Oxford Biology Reader No. 44, ed. J. J. Head. London: Oxford University Press.
O'Hara, J. F., & Henshaw, G. G. (1982). The preparation of protoplasts from potato and related *Solanum* species. In *Plant tissue culture 1982*, ed. A. Fujiwara, pp. 591–92. Tokyo: Japanese Association for Plant Tissue Culture.
Patnaik, G., Wilson, D., & Cocking, E. C. (1981). Importance of enzyme purification for increased plating efficiency and plant regeneration from single protoplasts of *Petunia parodii*. *Z. Pflanzenphysiol.* 102, 199–205.
Power, J. B., & Cocking, E. C. (1971). Fusion of plant protoplasts. *Sci. Prog.* 59, 181–98.
Power, J. B., Berry, S. F., Cocking, E. C., Evans, P. K., & Frears, S. (1976). Isolation, culture and regeneration of leaf protoplasts in the genus *Petunia*. *Plant Sci. Lett.* 7, 51–5.
Raj, B., & Herr, J. M. (1970). Isolation of protoplasts from the placental cells of *Lycopersicum pimpinellifolium*. *Exp. Cell Res.* 64, 479–80.

Shepard, J. F., & Totten, R. E. (1975). Isolation and regeneration of tobacco mesophyll cell protoplasts under low osmotic conditions. *Plant Physiol. 55*, 689–94.
Strange, R. N., Pippard, D. J., & Strobel, G. A. (1982). A protoplast assay for phytotoxic metabolites produced by *Phytophthera dresehsleri* in culture. *Physiol. Plant Pathol. 20*, 359–64.
Upadhya, M. D. (1975). Isolation and culture of mesophyll protoplasts of potato. *Potato Res. 1518*, 438–45.
Vardi, E., Spiegel-Roy, P., & Galun, E. (1975). Citrus cell culture: Isolation of protoplasts, plating densities, effect of mutagens and regeneration of embryos. *Plant Sci. Lett. 4*, 231–6.
Vasil, V., & Vasil, I. K. (1973). Growth and cell division in isolated plant protoplasts in microchambers. In *Protoplasts et fusion cellules somatiques végétables*, pp. 139–49. Paris: Colloq. Int. Centre Nat. Rech. Sci.
Wetter, L. R., & Constabel, F. (eds. ) (1982). *Plant tissue culture methods*, 2d rev. Ed. Saskatoon: National Research Council of Canada, Prairie Regional Laboratory.
Willison, J. H. M., & Cocking, E. C. (1972). The production of microfibrils at the surface of isolated tomato fruit protoplasts. *Protoplasma 75*, 397–403.

# 14

## Protoplast fusion and somatic hybridization

The preceding chapter describes the methods used for the isolation, purification, and culture of isolated protoplasts, and offers some insight into the way in which whole plants may be regenerated from single isolated protoplasts. The interest in protoplast fusion techniques is related to the prospect that wider crosses than are now possible by sexual means may be achieved with protoplast fusion. For example, some plants that show physical or chemical incompatibility in normal sexual crosses may be produced by the fusion of protoplasts obtained from two cultures of different species. It should be emphasized, however, that hybrid whole plants have been regenerated in a relatively small number of fusion systems, and there are no instances of the successful use of somatic hybridization in a plant-breeding program. In the early research work, protoplasts could be isolated from a small number of plants; the number of successful protoplast isolations increased following improvements in technique. Doubtless, the number of successful fusion experiments will rapidly increase after the techniques have been perfected.

The fusion of plant protoplasts is not a particularly new phenomenon; Kuster in 1909 described the process of random fusion in mechanically isolated protoplasts. When two or more isolated protoplasts are fused together, there is always a coalescence of the cytoplasms of the various protoplasts (Fig. 14.1). The nuclei of the fused protoplasts may fuse together or remain separate. Cells containing nonidentical nuclei are referred to as "heterokaryons" or "heterokaryocytes" (Mastrangelo, 1979). The fusion of nuclei in a binucleate heterokaryon results in the formation of a true hybrid protoplast or "synkaryocyte" (Constabel, 1978). The fusion of two protoplasts from the same culture results in a "homokaryon." Frequently genetic information is lost from one of the nuclei. If one nucleus completely disappears, the cytoplasms of the two parental protoplasts are still hybridized (Fig. 14.1), and the fusion product is known as a "cybrid" (cytoplasmic hybrid) or "heteroplast." Certain genetic factors are carried in the cytoplasmic inheritance system instead of in the nuclear genes (e.g., male sterility in some plants). The formation of cybrids, therefore, has application in a plant-breeding program.

Spontaneous fusion of protoplasts may occur, or they may be induced to fuse in the presence of "fusigenic agents." During enzymatic digestion of the cell wall, the protoplasts of contiguous cells may fuse together through their adjoining plasmodesmata.

Figure 14.1. Some fusion products resulting from protoplast culture. The fusion of protoplasts A and B results in a binucleate heterokaryon containing the cytoplasmic contents of the two original protoplasts. Fusion of the two nuclei results in a tetraploid hybrid cell or synkaryocyte. If one of the nuclei degenerates, a cybrid or heteroplast is produced.

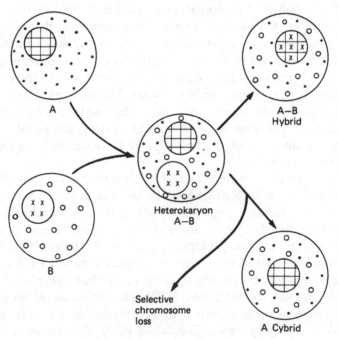

The dissolution of the wall allows the plasmodesmatal strands interconnecting the cells to enlarge; the cytoplasm and organelles from two or more cells then flow together. These spontaneous fusions are always intraspecific (i.e., originate from plant tissue of the same species). This phenomenon rarely occurs because the negative charges on the surface of the protoplasts cause them to repel each other (Grout & Coutts, 1974; Nagata & Melchers, 1978). Although the fusigenic agent lowers the surface charge, which permits the protoplast membranes to come into proximity, the adhesion of the protoplasts is insufficient to bring about fusion without molecular alterations in the bilayer structure of the plasma membranes (Ahkong et al., 1975a).

Several compounds have been shown to have a fusigenic effect on protoplasts. Workers at the University of Nottingham (Power, Cummins, & Cocking, 1970) showed that the addition of sodium nitrate to the culture medium induced fusion of root protoplasts from oat (*Avena sativum* L.) and maize (*Zea mays* L.). Fusion is also promoted by a combination of high pH (10.5), a high concentration of calcium ions (50 mM $CaCl_2 \cdot 2H_2O$), and high temperature (37 °C) (Keller & Melchers, 1973; Melchers & Labib, 1974). At present polyethylene glycol (PEG) is the most widely used fusigenic agent (Constabel & Kao, 1974; Kao & Michayluk, 1974). The molecular weight and concentration of PEG, the density of protoplasts, the incubation temperature, and the presence of divalent cations are all factors that play a role in the fusion process (Wallin, Glimelius, & Eriksson, 1974). Polyethylene glycol had been used to induce the fusion of animal cells (Pontecorvo, 1975); therefore, it was not surprising that a heterokaryon was produced between an animal cell and a plant cell (Ahkong et al., 1975b). The latter heterokaryon involved the fusion of a hen erythrocyte and a yeast protoplast. The fusion of cultured amphibian cells with protoplasts of a higher plant has also been reported (Davey et al., 1978).

The experiment outlined in this chapter involves an attempt to induce fusion between isolated protoplasts from two different plant sources. The inductive treatment, which involves low-speed centrifugation of the mixed culture in the presence of PEG, will result in a range of fusion products. The protoplast population will consist of unfused parental cells from the two tissues, homokaryons, heterokaryons, and multiple fusion products. If this mixture is plated on a culture medium, some of the various cell types will divide and develop callus. The next problem for the investigator is the recognition of callus formed by somatic hybrid and cybrid cells. The selection procedures are generally of two types: visual and biochemical.

Visual selection has been restricted to the fusion of colorless protoplasts with those containing chloroplasts. Protoplasts that demonstrate the completely integrated structural characteristics of both parental types are heterokaryons and potentially hybrid cells (Kao et al., 1974; Power & Cocking, 1977; Patnaik et al., 1982). An example of the fusion of plant protoplasts is shown in Figure 14.2. It is possible to attach a fluorescent label to the outer membranes of two parent protoplast populations and then separate the fusion products from the parental mixture by a method of fluorescent cell sorting (Redenbaugh et al., 1982).

Somatic hybrids can be selected by using a medium to encourage selective growth. A method was developed for the selection of hybrids resulting from the fusion of protoplasts of *Nicotiana glauca* and *N. langsdorffii* (Carlson, Smith, & Dearing, 1972) (Figure 14.3). Neither of the parental protoplasts are capable of growth on a medium deprived of auxin. The protoplast fusion products of the two species are auxin autotropic and are capable of callus formation on an auxin-free medium. The callus can then be subcultured and induced to regenerate hybrid plants. The rationale for this procedure was later confirmed by Smith, Kao, & Combatti (1976).

Figure 14.2. Fusion of colorless and chloroplast-containing protoplasts. (Courtesy of Professor G. Melchers.)

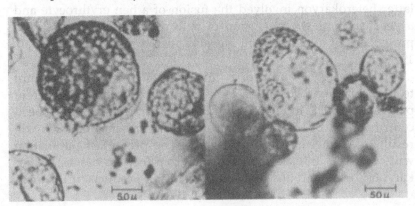

Another method involves the use of biochemical mutants and a selection of somatic hybrids by a form of complementation. The antibiotic actinomycin D was used in the detection of fusion products of two species of *Petunia* (Power et al., 1976). Cultured cells of *P. hybrida* cannot grow in the presence of actinomycin D, whereas cells of *P. parodii* are capable of growth in the presence of this antibiotic. The cells of the latter species, however, are unable to regenerate *Petunia* plants from callus cultures. The only cells capable of growth in the presence of actinomycin D and capable of regeneration of whole plants are the fusion products of the two parental protoplast lines. Two mutant strains of *Nicotiana tabacum*, which have the characteristics of light sensitivity and chlorophyll deficiency, were used in a complementation selection procedure by Melchers and Labib (1974) and Bottcher, Aviv, & Galun (1989).

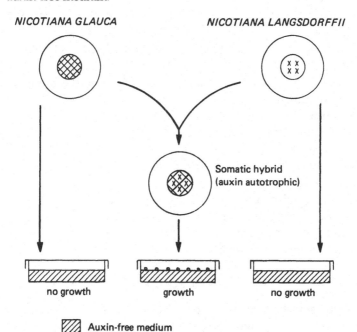

Figure 14.3. Screening method for the detection of somatic hybrids that are auxin autotropic in nutrition. The isolated protoplasts of both *Nicotiana glauca* and *N. langsdorffii* are unable to grow in the absence of exogenous auxin. The fusion product of the two parental types is auxin autotropic and grows on an auxin-free medium.

The techniques outlined in this chapter are probably among the most difficult that the student will encounter in this book. The fusion experiment should not be attempted until protoplasts have been successfully isolated and cultured (Chapter 13). The present experiment involves the induction of fusion of two protoplasts carrying distinct visual markers in the form of pigment color. The fusion products can be identified by microscopic examination of the cells. After this technique is mastered, the student can advance to more complex hybridization and selection methods.

### LIST OF MATERIALS

*Sterilization mode:* C, chemical; O, oven; A, autoclave

- isolated protoplasts of two types in culture medium (Chapter 13) from one of the following plants:
  a. leaves and tubers of *Beta vulgaris*
  b. leaves and taproot of *Daucus carota*
  c. leaves and petals of *Nicotiana tabacum*
  d. leaves and petals of *Petunia hybrida*

O  10-cm Petri dishes (10)
O  forceps
O  centrifuge tubes, 6×16 mm (capped with aluminum foil)
A  250-cm$^3$ Erlenmeyer flask with 200 cm$^3$ MS–mannitol (13% w/v) plus agar (2% w/v), 40 °C (Chapter 13, Procedure)
A  polyethylene glycol solution (20% w/v; relative molecular weight $M_r$ = 3,000)
A  Pasteur pipettes, wide bore
-  Fuchs–Rosenthal hemocytometer, 0.2-mm field depth
-  ethanol (80% v/v) dip
-  ethanol (70% v/v) in plastic squeeze bottle
-  methanol lamp
-  bench-top centrifuge, swinging-bucket type
-  Evans blue dye reagent
-  fluorescein diacetate
-  fluorescence microscopy

### PROCEDURE

1. After selection of the appropriate plants, isolate two sets of protoplasts by employing the technique outlined in the previous

chapter. This should result in one tube containing green protoplasts of mature leaf material and a second tube of red protoplasts from petal, tuber, or taproot tissue. Thus the markers for fusion are chloroplasts and vacuoles containing anthocyanin pigment.

2. The basic principle of fusion is shown in Figure 14.1. Similar numbers of protoplasts A and B are mixed in a centrifuge tube containing PEG (20% w/v) as the fusigenic agent. The tube is centrifuged at 75–100 $g$ for 10 min. This relatively slight pressure forces the protoplasts into close contact and allows fusion to take place.

3. The pellet of fused and unfused protoplasts is carefully resuspended and assayed for viability and density, and the mixture is plated out as described in Chapter 13.

## RESULTS

After the protoplast mixture has been plated and the agar has solidified, the plate may be viewed microscopically for the identification of the fusion products. Heterokaryons can be identified by the presence of chloroplasts and an anthocyanin-containing vacuole. There are many different combinations of fusion products, and the student should attempt to identify as many as possible. If any of the fusion products initiates a callus, plant regeneration can be attempted as previously described in Chapter 6.

## QUESTIONS FOR DISCUSSION

1. What chemical compounds have been employed as fusigenic agents?
2. What are the advantages of protoplast fusion over traditional methods of sexual hybridization?
3. What types of procedures can be used for the selection of hybrid cells?
4. What is a cybrid? How does this phenomenon occur, and does it have any significance in the breeding of plants?
5. List some interesting protoplast fusions that might result in unusual hybrids (e.g., potato and tomato).

## SELECTED REFERENCES

Ahkong, Q. F., Fisher, D., Tampion, W., & Lucy, J. A. (1975a). Mechanisms of cell fusion. *Nature 253*, 194–5.
Ahkong, Q. F., Howell, J. I., Lucy, J. A., Safwat, F., Davey, M. R., & Cock-

ing, E. C. (1975b). Fusion of hen erythrocytes with yeast protoplasts induced by polyethylene glycol. *Nature* 255, 66–7.

Bottcher, U. F., Aviv, D., & Galun, E. (1989). Complementation between protoplasts treated with either of two metabolic inhibitors results in somatic-hybrid plants. *Plant Science* 63, 67–77.

Carlson, P. S., Smith, H. H., & Dearing, R. D. (1972). Parasexual interspecific plant hybridisation. *Proc. Natl. Acad. Sci. USA* 69, 2292–4.

Constabel, F. (1978). Development of protoplast fusion products, heterokaryocytes, and hybrid cells. In *Frontiers of plant tissue culture 1978*, ed. T. A. Thorpe, pp. 141–9. Calgary: IAPTC.

Constabel, F., & Kao, K. N. (1974). Agglutination and fusion of plant protoplasts by polyethylene glycol. *Can. J. Bot.* 52, 1603–6.

Davey, M. R., Clothier, R. H., Balls, M., & Cocking, E. C. (1978). Ultrastructural study of fusion of cultured amphibian cells with higher plant protoplasts. *Protoplasma* 96, 157–72.

Grout, B. W., & Coutts, R. H. A. (1974). Additives for enhancement of fusion and endocytosis in higher plant protoplasts: An electrophoretic study. *Plant Sci. Lett.* 2, 397–403.

Kao, K. N., Constabel, F., Michayluk, M. R., & Gamborg, O. L. (1974). Plant protoplast fusion and growth of intergeneric hybrid cells. *Planta* 120, 215–27.

Kao, K. N., & Michayluk, M. R. (1974). A method for high frequency intergeneric fusion of plant protoplasts. *Planta* 115, 355–67.

Keller, W. A., & Melchers, G. (1973). Effect of high pH and calcium on tobacco leaf protoplast fusion. *Z. Naturforsch.* 28, 737–41.

Kuster, E. (1909). Über die Verschmelzung nackter Protoplasten. *Ber. Dtsch. Bot. Ges.* 27, 589–98.

Mastrangelo, I. A. (1979). Protoplast fusion and organelle transfer. In *Nicotiana: Procedures for experimental use*, ed. R. D. Durbin, pp. 65–73. Washington, D.C.: USDA Tech. Bull. No. 1586.

Melchers, G., & Labib, G. (1974). Somatic hybridisation of plants by fusion of protoplasts: I. Selection of light resistant hybrids of "haploid" light sensitive varieties of tobacco. *Mol. Gen. Genet.* 135, 277–94.

Nagata, T., & Melchers, G. (1978). Surface charge of protoplasts and their significance in cell–cell interactions. *Planta* 142, 235–8.

Patnaik, G., Cocking, E. C., Hamill, J., & Pental, D. (1982). A simple procedure for the manual isolation and identification of plant heterokaryons. *Plant Sci. Lett.* 24, 105–10.

Pontecorvo, G. (1975). Production of mammalian somatic cell hybrids by means of polyethylene glycol treatment. *Somatic Cell Genet.* 1, 397–400.

Power, J. B., & Cocking, E. C. (1977). Selection systems for somatic hybrids. In *Applied and fundamental aspects of plant cell, tissue, and organ culture*, ed. J. Reinert & Y. P. S. Bajaj, pp. 497–505. Berlin: Springer–Verlag.

Power, J. B., Cummins, S. E., & Cocking, E. C. (1970). Fusion of isolated plant protoplasts. *Nature* 225, 1016–18.

Power, J. B., Frearson, E. M., Hayward, C., George, D., Evans, P. K., Berry, S. F., & Cocking, E. C. (1976). Somatic hybridisation of *Petunia hybrida* and *Petunia parodii*. *Nature* 263, 500–2.

Redenbaugh, K., Ruzin, S., Bartholomew, J., & Basshan, J. (1982). Characterization and separation of plant protoplasts via flow cytometry and cell sorting. *Z. Pflanzenphysiol.* 107, 65–80.

Smith, H. H., Kao, K. N., & Combatti, N. C. (1976). Interspecific hybridisation by protoplast fusion in *Nicotiana*: Confirmation and extension. *J. Hered.* 67, 123–8.

Wallin, S., Glimelius, K., & Eriksson, T. (1974). The induction of aggregation and fusion of *Daucus carota* protoplasts by polyethylene glycol. *Z. Pflanzenphysiol.* 74, 64–80.

PART V

*Supplementary topics*

# 15

## Cryopreservation of germplasm

Considerable interest has been shown in recent years in the application of tissue culture technology to the storage of plant germplasm (Bajaj & Reinert, 1977; Wilkins & Dodds, 1983a,b). The conventional methods of germplasm preservation are prone to possible catastrophic losses because of:

1. attack by pests and pathogens,
2. climatic disorders,
3. natural disasters, and
4. political and economic causes.

In addition, the seeds of many important crop plants lose their viability in a short time under conventional storage systems. Examples of plants bearing short-lived seeds include *Theobroma cacao* (cocoa), *Artocarpus* spp. (breadfruit, jackfruit), *Persea americana* (avocado), and *Cocos nucifera* (coconut). Further reasons for using tissue culture procedures in the conservation of germplasm are also discussed in this chapter.

Vegetatively propagated plants that have a high degree of heterozygosity or that do not produce seed must be stored in a vegetative form as tubers, roots, or cuttings. This usually involves high-cost labor during the growing season and, in addition, the field collection and storage of propagules for short periods of time. Crops that fall into this category include yams (*Dioscorea* spp.), bananas and plantains (*Musa* spp.), potatoes (*Solanum* spp.), cassava (*Manihot*), taro (*Colocasia esculentum*), and sweet potato (*Ipomoea batatas*). An interesting example is the large-scale propagation of the international collection of potato germplasm at the International Potato Center (CIP), Peru, which in 1980 involved

12,000 accessions. Several of these potatoes are infertile polyploid cultivars for which seed storage is obviously impossible.

For many species, including bananas and plantains (*Musa* spp.), coconut (*Cocos nucifera*), oil palm (*Elaeis guineensis*), and date palm (*Phoenix dactylifera*), there is no efficient method of large-scale vegetative propagation. The conventional method of reproduction of date palm – that is, by suckers arising from the base of the main stem – is slow and unreliable. The development of a suitable tissue culture system of propagation and conservation of germplasm would therefore have obvious benefits. In the case of date palm the situation is especially serious. The genomes of this species in Algeria and Morocco are bordering on extinction because of Bayoun disease (*Fusarium oxysporum* Schlect. var. *albedinis*).

Many long-lived forest trees, including angiosperms and gymnosperms, do not produce seed until a certain age, and these trees must be propagated vegetatively when it is necessary to produce the parental genotype. The most frequently used methods of vegetative propagation of rosaceous fruit trees are budding and grafting. One area of research currently under investigation is the possibility of rapid clonal propagation of fruit trees by means of tissue culture techniques. These techniques have special relevance to certain fruit tree cultivars that are either difficult and/or expensive to propagate by conventional methods. For example, scions of apple cultivars are usually propagated by budding or grafting to rootstocks that are themselves raised by stooling or layering. This process takes three years and demands expensive nursery facilities and skills.

Propagation by the production of self-rooted plants would be much more rapid and could be of value in hastening the availability of new cultivars developed from breeding programs. An additional advantage of tissue culture techniques for fruit tree propagation is the rapid in vitro multiplication and rooting of apple rootstocks such as M9. This rootstock is widely used because of its effects on precocity and on the control of tree and fruit size; a major disadvantage, however, is that cuttings are extremely difficult to root using conventional methods. Micropropagation with in vitro techniques has also been applied to *Prunus* cultivars, thus making available large quantities of material for use in breeding programs (Jones, Pontikis, & Hopgood, 1979).

*Cryopreservation.* It will be evident from the previous discussion that an in vitro system with a high multiplication rate, although ideal for purposes of clonal propagation, is entirely unsuitable as a means of germplasm conservation. Such systems require frequent attention and maintenance, and in some cases, carry the risk of genetic instability (Scowcroft, 1984). This instability is related to growth, particularly disorganized growth such as in callus cultures. Consequently, an ideal system for germplasm storage would be to store material in such a manner as to achieve complete cessation of cell division and growth. This can be accomplished by storing the plant material at the temperature of liquid nitrogen (−196 °C). Although such techniques have been applied to a range of tissue cultures, the success rates have been variable. Some aspects of cryopreservation (freeze preservation) are discussed by Withers (1980, 1982a, 1983b, 1987, 1990, 1991, 1992).

*Minimal-growth storage.* Techniques of germplasm conservation based on the storage of shoot-tip cultures or meristem-derived plantlets under conditions that permit only minimal rates of growth will have widespread application in the near future. Such systems already have important uses in several international germplasm resource centers, mainly because the stored material is readily available for use, it can easily be seen to be alive, and the cultures may be readily replenished when necessary (Wilkins & Dodds, 1983b).

There have been several approaches to growth suppression in plant tissue cultures. Three principal methods are used:

1. The physical conditions of culture can be altered (e.g., temperature or the gas composition within the culture vessels).
2. The basal medium can be altered, for example, using sub- or supraoptimal concentrations of nutrients. Some factor essential for normal growth may either be omitted or be employed at a reduced level.
3. The medium can be supplemented with growth retardants (e.g., abscisic acid) or osmoregulatory compounds such as mannitol and sorbitol.

Reviews of tissue culture conservation via minimal-growth techniques have been given by several workers (Henshaw et al., 1980;

Withers, 1980). Several aspects of these techniques, with the potential for immediate application, are discussed here.

Temperature reduction has been very effective as a means of storage of tissue cultures. Some of the many species that can be conserved by this method include grapes (*Vitis rupestris*), potatoes (*Solanum*), beets (*Beta vulgaris*), apples (*Malus domestica*), strawberries (*Fragaria vesca*), sweet potato (*Ipomoea batatas*), cassava (*Manihot esculentum*), and various forage grasses (e.g., *Lolium*, *Festuca*, *Dactylis*, and *Phleum*). In all cases the storage temperature employed is dependent on the particular crop species. As a general rule, temperate crops, such as potatoes, strawberries, grasses, and apples, are stored at 0–6 °C; whereas tropical crops, such as cassava and sweet potato, are stored within the range 15–20 °C.

For certain species the use of a reduced-temperature storage system has proved very successful. Meristem-derived plantlets of strawberry have been stored for as long as six years. It is not surprising that at least three international genetic resource centers currently use this technique on a routine basis.

The use of a decreased atmospheric pressure and a lowered partial pressure of oxygen has been proposed as a means of minimal-growth storage (Bridgen & Staby, 1983). The addition of growth retardants to culture media as a means of inducing minimal growth has been attempted by several workers. The compounds employed have included abscisic acid, mannitol, sorbitol, tributyl-2,4-dichlorobenzylphosphonium chloride (Phosfon D), maleic hydrazide, succinic acid-2,2-dimethyl hydrazide (B-995), (2-chloroethyl)-trimethylammonium chloride (CCC), and ancymidol. The effects of some of these growth retardants on the survival of potato cultures were given by Westcott (1981). The survival of potato cultivars after 12 months' growth in the presence of these inhibitory substances were roughly equivalent to the results obtained by using reduced temperatures and decreased levels of nutrients (Westcott, 1981). A choice of storage methods is therefore available.

Investigations at the University of Birmingham have shown that cultures of temperate fruit trees may be stored for several months by various techniques. Cherry shoot cultures have survived on a liquid medium at normal temperatures for prolonged periods (Wilkins, Bengochea, & Dodds, 1982; Withers, 1991, 1992). Similar unpublished results have been obtained by this research group with cultures of apple, plum, and pear cultivars.

In the present student experiments shoot tips of potato will be conserved by freezing in liquid nitrogen and by culture on a medium containing maleic hydrazide as a growth inhibitor.

LIST OF MATERIALS

*Sterilization mode:* C, chemical; O, oven; A, autoclave

- shoot-apex culture of potato exhibiting regeneration (Chapter 10)
- C  scalpel
- O  forceps
- O  Petri dishes, 100 × 15 mm (five)
- O  culture tubes (30)
- O  aluminum foil (cut into squares)
- A  200 cm$^3$ potato shoot micropropagation medium (Chapter 10; sufficient amount for 20 culture tubes)
- A  100 cm$^3$ potato shoot micropropagation medium supplemented with maleic hydrazide (10 mg/l) (employ 10 culture tubes)
- ethanol (80% v/v) dip
- ethanol (70% v/v) in plastic squeeze bottle
- Dewar flask of liquid nitrogen
- face mask (protective plastic)
- asbestos gloves
- long forceps, approximately 30–40 cm in length
- plant growth chamber (illuminated; 25 °C)

PROCEDURE

*Experiment 1. Storage of shoots in the presence of growth retardant*

1. Actively propagating potato shoots (Chapter 10) are removed from the culture vessel and transferred to sterile Petri dishes for dissection of the individual shoots.

2. Using sterile instruments 10 shoots are transferred to culture tubes containing a fresh micropropagation medium as a control, and 10 shoots are transferred to tubes containing a similar medium supplemented with mannitol 6%.

Figure 15.1. Effect of the growth retardant maleic hydrazide on in vitro cultures. The left-hand tube shows freshly transferred culture; the one at the right shows control culture after six weeks. Intermediate tubes contain 10 and 1 mg/l of maleic hydrazide.

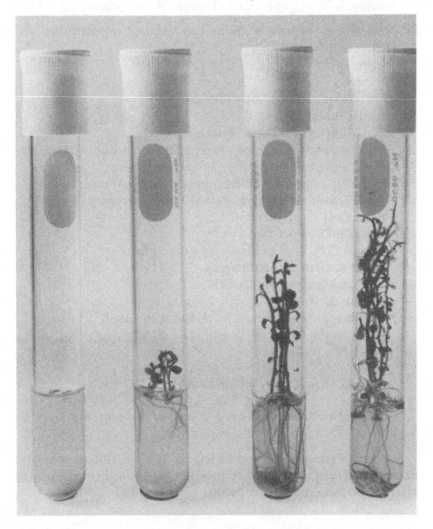

3. The cultures are incubated in illuminated plant growth chambers (25 °C). Growth measurements are made of the cultures at monthly intervals.

### Experiment 2. Freeze preservation of potato shoot tips

1. Excise some sterile potato shoot apices (Chapter 10). Wrap them carefully in sterile squares of aluminum foil.
2. While wearing protective gloves and a face mask, plunge a foil packet into a Dewar flask of liquid nitrogen (–196 °C). Permit the packet to remain in the liquid nitrogen for several minutes. The sample can be left for longer periods if a liquid nitrogen storage tank is available.
3. Remove the foil packet and allow it to warm to room temperature.
4. Open the packet and carefully transfer aseptically the frozen and thawed apices to culture tubes containing the potato shoot micropropagation medium (10 tubes).

## RESULTS

### Experiment 1. Storage of shoots in the presence of growth retardant

The control shoots transferred to the normal micropropagation medium grow rapidly under these culture conditions. The growth retardant has a severe effect on shoot growth. A comparison of growth retardant to control cultures following a six-week period is shown in Figure 15.1.

### Experiment 2. Freeze preservation of potato shoot tips

The time required for excised shoot apices to initiate growth varies greatly, but growth should be easily visible after four to six weeks with potato. The success rate or percentage of survivors will be relatively low, and results will vary from one species to another.

## QUESTIONS FOR DISCUSSION

1. Why are tissue cultures used to conserve germplasm of plants?
2. Name some plants in which tissue culture is the only method for germplasm conservation.
3. What techniques are available for in vitro conservation? Discuss the advantages and disadvantages of each technique.

## SELECTED REFERENCES

Bajaj, Y. P. S., & Reinert, J. (1977). Cryobiology of plant cell cultures, and establishment of gene banks. In *Applied and fundamental aspects of plant cell, tissue and organ culture,* ed. J. Reinert & Y. P. S. Bajaj, pp. 757–76. Berlin: Springer-Verlag.

Bridgen, M. P., & Staby, G. L. (1983). Protocols of low-pressure storage. In *Handbook of plant cell culture,* vol. 1, *Techniques for propagation and breeding,* ed. D. A. Evans, W. R. Sharp, P. V. Ammirato, & Y. Yamada, pp. 816–27. New York: Macmillan.

Henshaw, G. G., O'Hara, J. F., & Westcott, R. J. (1980). Tissue culture methods for the storage and utilization of potato germplasm. In *Tissue culture methods for plant pathologists,* ed. D. S. Ingram & J. P. Helgeson, pp. 71–6. Oxford: Blackwell Scientific Publications.

Jones, O. P., Pontikis, C. A., & Hopgood, M. E. (1979). Propagation in vitro of five apple scion cultivars. *J. Hort. Sci.* **54,** 155–8.

Scowcroft, W. R. (1984) *Genetic variability in tissue culture: Impact on germplasm conservation and utilisation.* Rome: IBPGR.

Westcott, R. J. (1981). Effect of growth retardants on growth of potato shoots in vitro. *Potato Res.* **24,** 331–7.

Wilkins, C. P., Bengochea, T., & Dodds, J. H. (1982). In vitro conservation of plant genetic resources. *Outlook Agric.* **11,** 67–73.

Wilkins, C. P., & Dodds, J. H. (1983a). The application of tissue culture techniques to plant genetic conservation. *Sci. Prog.* **68,** 259–84.

(1983b). Tissue culture conservation of woody species. In *Tissue culture of trees,* ed. J. H. Dodds, pp. 113–38. London: Croom Helm.

Withers, L. A. (1980). Tissue culture storage for genetic conservation. *IBPGR Tech. Report.* Rome: FAO, United Nations.

(1982a). Storage of plant tissue cultures. In *Crop genetic resources: The conservation of difficult material,* ed. L. A. Withers & J. T. Williams, pp. 49–82. Paris: International Union of Biological Sciences, IBPGR.

(1983b). Germplasm storage in plant biotechnology. In *Plant biotechnology,* ed. S. H. Mantell & H. Smith, pp. 187–218. Cambridge: Cambridge University Press.

(1987). Long-term preservation of plant cells, tissue and organs. In *Oxford Surv. Plant Mol. Cell Biol.* **4,** 221–72.

(1990). Cryopreservation of plant cells. In *Methods in molecular biology,*

vol. 6, *Plant cell and tissue culture*, ed. J. W. Pollard & J. M. Walker, pp. 39–48. Totowa, N.J.: The Humana Press.
(1991). Maintenance of plant tissue cultures. In *Maintenance of microorganisms*, 2d Ed., pp. 243–67. London: Academic Press.
(1992). Biotechnology of perennial fruit crops. In *Biotechnology of perennial fruit crops*, ed. F. A. Hammerschlag & R. E. Litz, pp. 169–200. C·A·B International.

# 16

## Production of secondary metabolites

Aside from the primary metabolic pathways common to all life forms, some reactions lead to the formation of compounds unique to a few species or even to a single cultivar. These reactions are classified under the term "secondary metabolism," and their products are known as "secondary metabolites" (Luckner & Nover, 1977). These substances include alkaloids, antibiotics, volatile oils, resins, tannins, cardiac glycosides, sterols, and saponins. In addition to their economic importance, many secondary metabolites play ecological and physiological roles in higher plants. Investigations in the area of biochemical ecology indicate that some secondary compounds produced by plants are important either to protect these plants against microorganisms and animals, or to enhance the ability of one plant species to compete with other plants in a particular habitat (Bell, 1980). For additional background, Mothes (1980) has given an excellent introduction to secondary plant products.

Despite advances in the field of organic chemistry, plants are still an important commercial source of chemical and medicinal compounds. The chief industrial applications of secondary metabolites have been as pharmaceuticals (e.g., sterols and alkaloids), and as agents in food flavoring and perfumery (Collin & Watts, 1983). In some cases, these plants have not been subjected to intensive genetic programs for the optimum production of the compound. In addition, there have been technical and economic problems in the cultivation of these plants. Unfortunately, many Third World countries producing medicinal plants are politically unstable, and the supply of crude plant material for processing cannot be guaranteed.

It has been proposed that many of these secondary metabolites produced by intact plants could be synthesized by cell cultures (Klein, 1960; Puhan & Martin, 1970). The basic technology involved in submerged cell cultures on a large scale was described by Nickell (1962). Patents have been obtained for production from cell cultures of such metabolites as allergens, diosgenin, L-dopa, ginseng saponin glycosides, and glycyrrhizin (Staba, 1977). Tissue cultures have produced compounds previously undescribed, and cultures of higher plant cells may provide an important source of new, economically important compounds (Butcher, 1977).

Although the production of secondary metabolites by cell cultures may be impractical, in some cases the techniques of plant tissue culture can be used to improve the cultivation of these plants. These culture procedures include vegetative propagation, the isolation of virus-free stock, mutation studies with haploid plants, protoplast fusion, and the screening of disease-resistant lines.

There are numerous reasons why progress has been slow in the industrial application of cell cultures for the production of secondary metabolites. The cultures exhibit relatively slow rates of growth, and the biosynthesis of the desired compounds is often at a much lower level than in the intact plant. In order for cell cultures to be used as commercial sources of these compounds, the in vitro production must be comparable to or exceed the amount produced by the intact plant. Several reports have been published indicating yields approaching or exceeding yields from the whole plant (Table 16.1). In some cases, the production of secondary metabolites does not show a positive correlation with the maximal growth rate of the culture. This observation may reflect a competition for metabolites utilized in primary metabolism with those pathways leading to the formation of secondary products; for example, competition could exist for amino acids in the formation of proteins, alkaloids, and phenylpropanoids (Aitchison, Macleod, & Yeoman, 1977).

One approach used to regulate metabolic pathways favoring the production of secondary metabolites has been to add precursors to the culture medium. Although enrichment of the medium with precursors has given some degree of success, failure of the desire results may be due to lack of uptake, precipitation, conjugation, diversion into alternative pathways, or lack of one of the enzymes between precursor and product (Aitchison et al., 1977). Attempts have been made to increase the yields of cell cultures by selecting high-yield strains (Tabata et al., 1978).

Although the biosynthesis of secondary metabolites is not accompanied by visible markers, radioimmunoscreening methods have been developed for cell cultures producing secondary metabolites (Weiler, 1977). The instability of cell cultures for the continued production of secondary products poses a problem since some cell lines lose the ability to synthesize the desired compound after prolonged culture. Obviously, it is important to examine thoroughly the continuing production of a given strain before scaling up for industrial production (Alfermann & Reinhard, 1978). Finally, there is the high cost involved in the large-scale production of cell cultures (e.g., in glassware, chemicals, and technicians). With the present level of development it appears that only rare and very expensive chemicals can be commercially produced by this method (Zenk, 1978).

The relationship between the degree of tissue organization and the biosynthesis of secondary products is obscure. The spatial orientation of enzymes, compartmentalization of enzymes and substrates, and reservoir sites for product accumulation may be some of the factors involved in the biosynthesis of secondary products by specialized tissues (Butcher, 1977). The metabolic requirements

Table 16.1. *Secondary metabolites produced in cell cultures at levels equal to or higher than those found in the intact plant*

| Compound | Plant species | Reference |
|---|---|---|
| nicotine | *Nicotiana rustica* | Tabata & Hiraoka, 1976 |
| serpentine | *Catharanthus roseus* | Döller, Alfermann, & Reinhard, 1976 |
| anthaquinones | *Morinda citrifolia* | Zenk, El-Shagi, & Schulte, 1975 |
| diosgenin | *Dioscorea deltoidea* | Kaul, Stohs, & Staba, 1969 |
| thebain | *Papaver bracteatum* | Kamimura, Akutsu, & Nishikawa, 1976 |
| proteinase inhibitors | *Scopolia japonica* | Misawa et al., 1975 |
| ginseng saponins | *Panax ginseng* | Jhang, Staba, & Kim, 1974 |
| phenolics | *Acer pseudoplatanus* | Westcott & Henshaw, 1976 |
| flavonols | *Acer pseudoplatanus* | Westcott & Henshaw, 1976 |
| coumarin derivatives | *Ruta graveolens* | Steck et al., 1971 |
| alkaloids | *Ruta graveolens* | Steck et al., 1971 |
| solasadine | *Solanum lacinatium* | Chandler & Dodds, 1983a,b |

for some of these biosynthetic pathways, however, do not depend on the level of cytodifferentiation. In a review on secondary-product formation in cell cultures, Butcher (1977) has subdivided these compounds into four general groups.

1. Some compounds occur throughout the plant kingdom and are not associated with any level of cytodifferentiation (e.g., phytosterols and certain flavonoids).
2. Some widely distributed compounds are restricted to certain types (e.g., lignins and tannins).
3. Some compounds are restricted to certain plant families and species, although the biosynthesis is not associated with any form of cytodifferentiation (e.g., specific flavonoids and anthraquinones).
4. The biosynthesis of some compounds is restricted to highly specialized cells or tissues (e.g., essential oils, resins, and latex). Within this group the level of differentiation is directly related to the biosynthesis of the compound.

In the last category, we can assume that progress toward inducing certain levels of cytodifferentiation in cell cultures must be made before success will be achieved in the in vitro biosynthesis of these secondary metabolites (Street, 1977; Yeoman et al., 1982).

Discussions of the possible relationships between secondary metabolism and cytodifferentiation have been published (Böhm, 1977; Yeoman et al., 1982). The main problem in establishing these relationships, as in most cell culture research, is that the method of culture is empirical and cannot be transferred from one plant species or variety to another.

Because the principles of chemical biosynthesis by cell cultures are poorly understood, we cannot anticipate a quick success as far as practical applications are concerned (Alfermann & Reinhard, 1978). Yet, in spite of numerous problems, there are active research programs on the in vitro production of secondary metabolites, particularly in Ireland, Japan, West Germany, Israel, India, and the United States.

Several secondary metabolites produced by cell cultures are pigmented (Reinert, Clauss, & Ardenne, 1964; Alfermann & Reinhard, 1971; Strickland & Sunderland, 1972). In this chapter the student will study the in vitro biosynthesis of secondary metabolites either by a visual examination of cultures for anthocyanin pigment formation or by the detection of aromatic compounds in mint cultures (Lin & Staba, 1961). The aromatic monoterpenes produced by mint can be detected by smelling the cultures or by gas–liquid chromatography, if available.

### LIST OF MATERIALS

*Sterilization mode:* C, chemical; O, oven; A, autoclave

- mature plants of *Mentha, Haplopappus,* or *Daucus carota* 300 $cm^3$ aqueous solution (10% v/v) commercial bleach containing 2 $cm^3$ of a wetting agent (1% v/v)
- paring knife
- C scalpel
- O forceps
- O 9-cm Petri dishes for explant culture (12)
- O 9-cm Petri dishes for explant preparation and rinsing (five)
- O stainless steel cork borer (No. 2) containing metal rod
- A 9-cm Petri dishes containing two sheets of Whatman No. 1 filter paper (five)
- A explant cutting guide
- A 400 $cm^3$ MS medium for callus induction (Chapters 4 and 5)
- A 125-$cm^3$ Erlenmeyer flasks containing 100 $cm^3$ $DDH_2O$ (12)
- O 600-$cm^3$ beakers (four)
- ethanol (80% v/v) dip
- ethanol (70% v/v) in plastic squeeze bottle
- methanol lamp
- interval timer
- heavy-duty aluminum foil (one roll)
- incubator (26–28 °C)

Leaf or stem material of *Mentha* can be employed for the study of aromatic compounds. Stem material of *Haplopappus* and taproot of *Daucus carota* will yield explants suitable for the production of anthocyanin pigments.

## PROCEDURE

1. Suitable plant material is surface sterilized according to the procedure given in Chapter 5. Explants of the tissue are rinsed in $DDH_2O$, blotted on Whatman No. 1 filter paper, and placed on the culture medium in order to initiate callus. The cultures are incubated at 26–28 °C in the presence or absence of artificial light.
2. After three weeks the callus may be subcultured to a freshly prepared medium.

*Anthocyanin pigment formation.* Following development of callus from the *Haplopappus* or *Daucus* explants, some areas of the callus will be white and other areas will be pigmented. Fragments of callus exhibiting varying degrees of pigmentation should be carefully excised from the callus mass and subcultured separately on a fresh medium. In this way it is possible to build up a number of "cell lines" with different anthocyanin pigment characteristics. This selection process is time consuming, and a minimum of several months will be required in order to develop a collection of cell lines.

*Aromatic terpenes.* After callus has been initiated from the *Mentha* explant, it should be subcultured on a fresh medium. Small portions of the callus can be "sniffed" to see if they still possess the characteristic mint odor.

If gas–liquid chromatography is available, transfer a fragment of callus to a culture tube sealed with a gas-tight fitting. After a few hours remove a gas sample from the culture tube with a hypodermic syringe. Apply the gas sample to the GLC instrument (flame ionization detector; 6 ft × 1/8 in., 15% DEGS-filled glass column; 40 $cm^3$/min $N_2$ and $H_2$ flow rate; injector, detector, and column temperatures adjusted to 180°, 180°, and 130 °C, respectively.) For additional information, see the procedure outlined by Aviv and Galun (1978).

## RESULTS

*Anthocyanin pigments.* As callus is initiated from the *Haplopappus* and *Daucus* explants, localized areas of pigmentation occur.

Some of these areas are completely white, whereas others have an orange or pink coloration. When small inocula are removed from the callus mass and subcultured, each cell line maintains its own distinctive coloration.

*Aromatic terpenes.* In the early stages of callus formation and during the first few subcultures, the odor of the callus clearly indicates the presence of the characteristic monoterpenes. As the callus is repeatedly subcultured, the levels of aromatic compounds gradually decline.

### QUESTIONS FOR DISCUSSION

1. Name some important pharmaceutical chemicals produced by plants.
2. What are the advantages of producing pharmaceutical compounds from cell and tissue cultures?
3. What is the genetic significance of cell line production of anthocyanin pigments? Are the cell lines genetically comparable?

### APPENDIX

Two additional experiments can be conducted on the production of secondary metabolites by callus cultures.

1. Subculture callus to an MS medium in which the mineral salts have been diluted by a factor of 10 (i.e., 1 part MS salts : 9 parts $DDH_2O$). In comparison with the original experiment, is the concentration of secondary products higher or lower?
2. Allow the callus culture to be maintained on the same medium without subculturing for approximately six weeks. Does prolonged culture on the same medium influence the production of secondary products?

### SELECTED REFERENCES

Aitchison, P. A., Macleod, A. J., & Yeoman, M. M. (1977). Growth patterns in tissue (callus) cultures. In *Plant tissue and cell culture,* ed. H. E. Street, pp. 267–306. Oxford: Blackwell Scientific Publications.

Alfermann, A. W., & Reinhard, E. (1971). Isolation of anthocyanin-producing and non-producing cell lines of tissue cultures of *Daucus carota. Experientia* 27, 353–4.

(1978). Possibilities and problems in production of natural compounds by cell culture methods. In *Production of natural compounds by cell culture methods*, ed. A. W. Alfermann & E. Reinhard, pp. 3–15. Munich: Gesellschaft für Strahlen und Umweltforschung.

Aviv, D., & Galun, E. (1978). Conversion of pulegone to isomenthone by cell suspension lines of *Mentha* chemotypes. In *Production of natural compounds by cell culture methods*, ed. A. W. Alfermann & E. Reinhard, pp. 60–7. Munich: Gesellschaft für Strahlen und Umweltforschung.

Bell, E. A. (1980). The possible significance of secondary compounds in plants. In *Encyclopedia of plant physiology*, n.s., vol. 8, *Secondary plant products*, ed. E. A. Bell & B. V. Charlwood, pp. 11–21. Berlin: Springer–Verlag.

Böhm, H. (1977). Secondary metabolism in cell cultures of higher plants and problems of differentiation. In *Secondary metabolism and cell differentiation*, ed. M. Luckner, L. Nover, & H. Böhm, pp. 104–23. Berlin: Springer–Verlag.

Butcher, D. N. (1977). Secondary products in tissue cultures. In *Applied and fundamental aspects of plant cell, tissue and organ culture*, ed. J. Reinert & Y. P. S. Bajaj, pp. 668–93. Berlin: Springer–Verlag.

Chandler, S. F., & Dodds, J. H. (1983a). The effect of phosphate, nitrogen and sucrose on the production of phenolics and solasidine in callus cultures of *Solanum lacinatum*. *Plant Cell Reports* 2, 205–8.

(1983b). Adventitious shoot initiation, in serially subcultured callus cultures of *Solanum lacinatum*. *Z. Pflanzenphysiol. 111*, 115–21.

Collin, H. A., & Watts, M. (1983). Flavor production in culture. In *Handbook of plant cell culture*, vol. 1, *Techniques for propagation and breeding*, ed. D. A. Evans, W. R. Sharp, P. V. Ammirato, & Y. Yamada, pp. 729–47. New York: Macmillan.

Döller, von G., Alfermann, A. W., & Reinhard, E. (1976). Production of indole alkaloids in tissue cultures of *Catharanthus roseus*. *Planta Med. 30*, 14–20.

Jhang, J. J., Staba, E. J., & Kim, J. Y. (1974). American and Korean ginseng tissue cultures: Growth, chemical analysis and plantlet production. *In Vitro 9*, 253–9.

Kamimura, S., Akutsu, M., & Nishikawa, M. (1976). Formation of thebain in the suspension culture of *Papaver bracteatum*. *Agric. Biol. Chem. 40*, 913–19.

Kaul, B., Stohs, S. J., & Staba, E. J. (1969). *Dioscorea* tissue cultures: III. Influence of various factors of diosgenin production by *Dioscorea deltoidea* callus and suspension cultures. *Lloydia 32*, 347–59.

Klein, R. M. (1960). Plant tissue cultures, a possible source of plant constituents. *Econ. Bot. 14*, 286–9.

Lin, M. L., & Staba, E. J. (1961). Peppermint and spearmint tissue cultures: I. Callus formation and submerged culture. *Lloydia 24*, 139–45.

Luckner, M., & Nover, L. (1977). Expression of secondary metabolism: An aspect of cell specialization of microorganisms, higher plants, and animals. In *Secondary metabolism and cell differentiation*, ed. M. Luckner, L. Nover, & H. Böhm, pp. 1–102. Berlin: Springer–Verlag.

Misawa, M., Tanaka, H., Chiyo, O., & Mukai, N. (1975). Production of plasmin inhibitory substances by *Scopolia japonica* suspension cultures. *Biotechnol. Bioeng.* 17, 305–14.

Mothes, K. (1980). Historical introduction. In *Encyclopedia of plant physiology*, n.s., vol. 8, *Secondary products*, ed. E. A. Bell & B. V. Charlwood, pp. 1–10. Berlin: Springer–Verlag.

Nickell, L. G. (1962). Submerged growth of plant cells. *Adv. Appl. Microbiol.* 4, 213–36.

Puhan, Z., & Martin, S. M. (1970). The industrial potential of plant cell culture. *Prog. Ind. Microbiol.* 9, 13–39.

Reinert, J., Clauss, H., & Ardenne, R. V. (1964). Anthocyanbildung in Gewebekulturen von *Haplopappus gracilis* in Licht verschiedener Qualität. *Naturwissenshaften* 51, 87.

Staba, E. J. (1977). Tissue culture and pharmacy. In *Applied and fundamental aspects of plant cell, tissue, and organ culture*, ed. J. Reinert & Y. P. S. Bajaj, pp. 694–702. Berlin: Springer–Verlag.

Steck, W., Bailey, B. K., Shyluk, J. P., & Gamborg, O. L. (1971). Coumarins and alkaloids from cell cultures of *Ruta graveolens*. *Phytochemistry (Oxf.)* 10, 191–8.

Street, H. E. (1977). Applications of cell suspension cultures. In *Applied and fundamental aspects of plant cell, tissue, and organ culture*, ed. J. Reinert & Y. P. S. Bajaj, pp. 649–67. Berlin: Springer–Verlag.

Strickland, R. G., & Sunderland, N. (1972). Production of anthocyanins, flavonals and chlorogenic acids by cultured callus tissues of *Haplopappus gracilis*. *Ann. Bot. (Lond.)* 36, 443–57.

Tabata, M., & Hiraoka, N. (1976). Variation of alkaloid production in *Nicotiana rustica* callus cultures. *Physiol. Plant.* 38, 19–23.

Tabata, M., Ogino, T, Yoshioka, K., Yoshikawa, N., & Hiraoka, N. (1978). Selection of cell lines with higher yields of secondary products. In *Frontiers of plant tissue culture 1978*, ed. T. A. Thorpe, pp. 213–22. Calgary: IAPTC.

Weiler, E. W. (1977). Radioimmuno-screening methods for secondary plant products. In *Plant tissue culture and its bio-technological application*, ed. W. Barz, E. Reinhard, & M. H. Zenk, pp. 266–77. Berlin: Springer–Verlag.

Westcott, R. J., & Henshaw, G. G. (1976). Phenolic synthesis and phenylalanine ammonia-lyase activity in suspension cultures of *Acer pseudoplatanus* L. *Planta* 131, 67–73.

Yeoman, M. M., Lindsey, K., Miedzybrodzka, M. B., & McLauchlan, W. R. (1982). Accumulation of secondary products as a facet of differentiation in plant cell and tissue cultures. In *Differentiation in vitro*, ed. M. M. Yeoman & D. E. S. Truman, pp. 65–82. Cambridge: Cambridge University Press.

Zenk, M. H. (1978). The impact of plant cell culture on industry. In *Frontiers of plant tissue culture 1978*, ed. T. A. Thorpe, pp. 1–13. Calgary: IAPTC.

Zenk, M. H., El-Shagi, H., & Schulte, U. (1975). Anthraquinine production by cell suspension cultures of *Morinda citrifolia*. *Plant Med.* 29, 79–101.

# 17

## Quantitation of procedures

The results of experiments conducted on plant tissue cultures are expressed in qualitative descriptions as well as in measurements of a quantitative nature. The growth of a culture over a period of time, whether it is a callus or a suspension culture, is characterized by an increase in cell number, an increase in volume or mass, and changes in biochemistry and cellular complexity.

Sorokin (1973), in a discussion of the terminology of growth in connection with algal suspensions, raised some points that can be applied to suspension cultures of higher plant cells. Because the term "growth" is ambiguous, referring both to the product of the process and to the process itself, it is preferable to designate the product as "yield," and reserve "growth" for the process of accumulation of the product. Yield does not represent "total growth" of the system, but only the net result of the metabolic processes (anabolic gain, catabolic loss). There will be losses also due to cell death. The expression of yield must be accompanied by the time period involved. In the case of batch cultures of cell suspensions, yield is determined at the cessation of growth, which is the beginning of the stationary phase.

A wide choice of parameters can be employed in the analysis of growth rates. In this brief introduction to quantitative analyses, we consider a few of the basic techniques of measurement that may be applied to root, callus, and suspension cultures (Table 17.1).

The simplest measurement involves the increase in length of cultured roots (see Chapter 9). Periodic measurements can be made with the metric rule held beneath the growing roots in the culture flasks or by the use of graph paper placed beneath the cultures. Sterility of the cultures is not endangered. Because cultured roots normally do not exhibit secondary thickening, the cross-sectional area remains constant during growth; thus, increments in length represent an accurate measure of volume change with growth (White, 1963). Although this method ignores branch-root formation, the error introduced in measurements over a seven-day period is never above 10% and seldom exceeds 1% (White, 1943).

Although the growth rates of callus cultures are frequently expressed on the basis of increase in fresh weight, these values should not be used in the estimation of cell numbers and cell division rates. Callus is heterogeneous, and a callus mass has centers of high division rates and regions of low metabolic activity. It is advantageous to measure fresh weight as a growth parameter because this is a rapid method of following an increase in tissue mass. Although it is possible to weigh callus samples periodically under aseptic conditions, the student will find that this practice almost invariably leads to some degree of contamination. The fresh weight of a cell suspension can be determined by collecting the cells on a filter of industrial nylon mesh. Directions for this procedure are given at the end of the chapter.

Measurement of the dry weight of a callus gives an acceptable estimation of the biosynthetic activity of a culture; moreover, at fresh weights below 500 mg, the relationship between fresh and dry weight is approximately linear (Wetter & Constabel, 1982).

Table 17.1. *Various techniques for the measurement of growth and differentiation in plant tissue cultures*

| Technique | Application[a] |
|---|---|
| linear growth rate | R |
| fresh weight | C, S, R* |
| dry weight | C, S, R* |
| density cell count | S |
| total cell count | C |
| packed cell volume | S |
| mitotic index | S, R, C* |
| tracheary element count | C, S* |

[a]The applications refer to cultures of root (R), callus (C), and suspensions (S). An asterisk indicates that the technique applies to a lesser extent.

One possible difficulty is the accumulation of large amounts of carbohydrates within the cells, which tends to complicate the interpretation of the data (Yeoman & Macleod, 1977). The dry weight of the callus at the termination of the experiment minus the dry weight of the primary explant at the beginning of the experiment, divided by the experimental period, gives the increase in dry weight per unit of time. Obviously, "the dry weight of the explant" does not refer to the explant that was actually cultured! A mean dry weight is obtained from several dummy explants of the same dimensions as the cultured explants. Drying the sample should be continued until two successive weighings of the sample and tray give a constant weight; prolonged drying will cause oxidative changes in cell dry weight in addition to water loss. The dry weight of a cell suspension can be determined by collecting the cells on a preweighed nylon filter. Cell suspension fresh and dry weights are normally given per cubic centimeter of culture and per $10^6$ cells (Street, 1977).

Another procedure used in connection with suspension cultures measures the total amount of cell mass in a given volume of culture by compressing the cells to a constant volume with centrifugation. The result is known as the packed cell volume (PCV). A known volume of the suspension is transferred to a graduated conical centrifuge tube. For greater accuracy, a special centrifuge tube terminating in a calibrated capillary tube is recommended. The latter vessel, employed in the measurement of the PCV of algal cultures, accepts a 4–7-$cm^3$ sample and has a capacity of 50 $\mu l$ in the calibrated section of the tube. The scale is subdivided into units of 1.0 $\mu l$, and estimations can be made to 0.1 $\mu l$ (Sorokin, 1973). Centrifugation conditions may vary with different cultures. Although Street (1977) recommends a centrifugation of 2,000 $g$ for 5 min for suspension cultures, a force of 1,500 $g$ for 30 min is recommended to pack a *Chlorella* sample to constant volume (Sorokin, 1973). Optimum conditions can be determined by taking a reading after a trial run, giving the sample an additional period of centrifugation, and making a second reading: The two readings should be the same. The main sources of error in the determination of PCV are

1. poor sampling of the original culture,
2. inadequate centrifugation, and

3. a delay in reading PCV, which leads to swelling of the cells and inflated values (Sorokin, 1973).

One important indication of growth is the increase in cell number of the culture. In agar estimates can be made of the total cell number of the primary explant and accompanying callus. The cell count in suspension cultures is expressed as density or concentration of cells per cubic centimeter of culture. The main problem with cell-counting methods is the separation of the cells into a homogeneous, unicellular sample. The approach is largely empirical because cultures differ widely in their response to cell-separation treatments. Many laboratories have used some modification of the Brown and Rickless (1949) technique of macerating cells with chromium trioxide or a mixture of chromium trioxide and hydrochloric acid. (An aqueous solution of chromium trioxide is chromic acid.) Although this is probably the best method of cell separation, care must be taken to avoid prolonged treatment leading to cell destruction. Some cultured tissues, however, will separate into individual cells following treatment with EDTA (Phillips & Dodds, 1977) or pectinase (Street, 1977). Following separation, a known volume of the sample is placed on a hemocytometer slide or a Sedgwick–Rafter slide for counting. A mean of 10 field counts (100× magnification) is obtained, and the total number of cells in the sample is then calculated. A discussion of the statistics of sampling phytoplankton with the Sedgwick–Rafter chamber is given by McAlice (1971).

The phenomenon of xylogenesis can be quantitated in an explant by counting the number of tracheary elements in a sample following cell separation by a suitable technique. The details of this method are given in the Procedure section of this chapter.

The mitotic index (MI) represents the percentage of the total cell population of a culture that at a given time exhibits some stage of mitosis. Rapidly growing cell populations exhibit an MI of approximately 3–5% (Wetter & Constabel, 1982). Although the MI is a useful technique for studying the growth of a culture, many factors influence its value: Among these are the time required for the completion of a cell cycle, the duration of mitosis, the percentage of noncycling and dead cells, and the degree of division synchrony of the cell population. Because of the involvement of these factors, the determination of the MI alone is not an accurate reflection of the division synchrony of a given culture. A better approach for appraising a culture's degree of division synchrony is to combine data from three sources:

1. measurement of the MI,
2. cell numbers showing the extent of synchrony of cytokinesis, and
3. degree of S-phase synchrony with tritiated thymidine.

Several studies have been made on the synchrony of division in actively dividing callus and suspension cultures involving the calculation of the MI (Yeoman, Evans, & Naik, 1966; Yeoman & Evans, 1967; Street, 1968). The procedure is tedious because at least 500 nuclei should be examined for signs of mitosis. The number of samples prepared for each callus growth is empirical, and it is impractical to measure the MI of large fragments of callus. In mature callus a relatively small proportion of the cells will be actively dividing, and these meristematic cells will be localized in small growth centers or cambial zones. The sampling is more suitable for cell suspensions than callus. Of the several nuclear staining methods that have been used, the present experiment will employ the carbol-fuchsin stain of Carr and Walker (1961) as modified for plant cells (Kao, 1982). Alternative staining procedures are given in Street (1977) and Yeoman and Macleod (1977).

## LIST OF MATERIALS

*Experiments 1, 2. Determination of fresh and dry weight*

- equipment for initiation of callus (Chapter 5)
- fresh plant material for the preparation of noncultured explants
- aluminum-foil weighing trays; if dry-weight determinations are to be made with cell suspension cultures, nylon mesh, cut into disks, will be necessary to collect the cells
- 9-cm Petri dishes, each containing two sheets of Whatman No. 1 filter paper (five)
- desiccator containing silica gel
- analytical balance
- drying oven (60 °C)

*Experiment 3. Determination of culture density by cell count*

suspension culture (Chapter 7)
aqueous solution (8% w/v) chromium trioxide
laboratory oven (70 °C)
bench-top shaker
vials for maceration; Pasteur pipettes
Sedgwick–Rafter chamber or hemocytometer slide
Whipple disk
stage micrometer slide
compound microscope (100× magnification)
microscope slides; cover slips
calculator, hand model
hand tally (optional)

*Experiment 4. Determination of packed cell volume (PCV)*

suspension culture (Chapter 7)
bench-top centrifuge (set at 2,000 $g$)
1-cm$^3$ pipettes
graduated conical centrifuge tubes, preferably terminating in a graduated capillary tube

*Experiment 5. Determination of mitotic index (MI)*

suspension culture (Chapter 7)
basic fuchsin
ethanol (70% v/v)
ethanol (95%)
phenol (5% w/v)
acetic acid (45% v/v)
glacial acetic acid
formaldehyde (37% v/v)
sorbitol
5-cm pipette, wide mouth
Pasteur pipettes microscope slides; cover slips; glass rod
lens paper
10-cm$^3$ vials, with caps
compound microscope (100× magnification)
hand tally (optional)

*Experiment 6. Tracheary element counts in primary explants*

- primary explants cultured on xylogenic medium (Chapter 12)
- maceration reagent consisting of 1:1 mixture of chromium (5% w/v) and hydrochloric acid (5% v/v)
- Pasteur pipettes
- 10-$cm^3$ vials, with caps
- syringe; No. 22-gauge and No. 18-gauge needles
- volumetric tube, 2-$cm^3$ calibration mark
- Sedgwich–Rafter chamber, with cover slip
- Whipple disk
- compound microscope (100× magnification); green filter (optional)
- calculator, hand model
- hand tally (optional)

PROCEDURE

*Experiment 1. Determination of fresh weight*

1. Select suitable plant material for the induction of callus and follow the instructions given in Chapter 5 (Procedure). For example, lettuce pith, taproot of carrot, or soybean cotyledon would be ideal material for this experiment. It is suggested that the same explants be used for both Experiments 1 and 2; that is, after determination of the fresh weight of each explant, the same explant can then be oven dried. A total of 30 explants will be required (10 explants for each of the three sampling periods).

2. Either before or after the culture procedure, prepare at least 10–15 additional dummy explants identical to the explants for culture. These noncultured explants need not be prepared under aseptic conditions.

3. After a brief rinse in distilled water, place the noncultured explants on a double layer of Whatman No. 1 filter paper in a Petri dish and cover it to prevent excessive water loss in the explants.

4. Weigh the blotted explants individually on a preweighed aluminum tray. They should be weighed on an analytical balance to the nearest 0.1 mg. Obtain a mean value for the gross weight (explant plus tray). Finally, subtract the weight of the tray from

the mean value of the noncultured explants. These noncultured explants can be oven dried and the determinations used in Experiment 2. (See the directions given in Experiment 2.)

5. Weigh and sacrifice 10 cultured explants exhibiting callus growth after varying time periods (e.g., 7, 14, and 21 days). Transfer the callus samples from the culture medium to a Petri dish containing filter paper before placing them in the weighing tray. Remove any traces of the gel medium that may be clinging to the underside of each sample. Carefully wipe the surface of the weighing tray after each weighing to remove traces of residual moisture. Handle the tray with forceps because moisture from the fingers will alter the weight.

6. After obtaining the gross fresh weight (sample plus tray), subtract the weight of the tray from the net weight of the sample. Obtain a mean value of the fresh weights of the 10 samples. Subtract the mean value of the noncultured explants from the mean value of the sample fresh weigh for each of the time periods (7-, 14-, and 21-day intervals). The latter values represent the fresh weight gain.

7. Plot the mean gain in fresh weight as a function of time.

## Experiment 2. Determination of dry weight

The explants cultured in Experiment 1 can be used for the dry-weight determination in this experiment. If an additional culture is necessary for this experiment, use the same plant material and culture medium as employed in Experiment 1. A comparison of the fresh- and dry-weight increments, under the same cultural conditions, can then be made.

1. The directions for this experiment parallel, to some extent, those for Experiment 1. Explants, cultured on a callus-inducing medium, are sacrificed at 7-, 14-, and 21-day intervals (10 explants for each time period). Place each callus sample in a preweighed aluminum-foil weighing tray. Then place the tray and sample in a Petri dish and dry at 60 °C for 12 hr. Cool the sample to room temperature in a desiccator containing silica gel, and weigh the sample (plus tray). Return the sample to the drying oven for an additional 4 hr, and repeat the process. If the sample was dried to constant weight, the two weighings will be the same. Subtract the dry weight of the tray from the preceding values to obtain the dry

weight of the sample alone. Obtain a mean value of the 10 samples taken at each time interval.

2. The noncultured explants prepared in Experiment 1 are oven dried and weighed in the same manner as the callus samples. In a similar manner the explants must be reweighed until a constant dry weight is obtained. Determine the mean value of the dry weight of the 10 samples of noncultured explants. Subtract the latter value from the mean of the callus-sample dry weight for each of the time periods. These values represent the gain in dry weight after 7, 14, and 21 days of growth.

3. Plot the mean gain in dry weight as a function of time on the same graph prepared for the fresh weight in Experiment 1. Are the fresh- and dry-weight increments parallel over the 21 days of culture?

### Experiment 3. Determination of culture density by cell count

This and the next experiment are devised to measure the growth rate of a suspension culture by determining the density or number of cells per cubic centimeter of culture (Experiment 3) and the packed cell volume (PCV) (Experiment 4). A density cell count was conducted earlier in preparing the inoculum for a suspension culture (Chapter 7, Procedure).

1. Prepare in advance a suitable suspension culture as outlined in Chapter 7. The culture should be in a state of active cell division and consist primarily of single cells and small clumps, with a relatively low level of lignification.

2. In order to separate the cell aggregates it is necessary to treat a sample of the suspension culture with chromium trioxide. It is suggested that the initial trial involve a mixture consisting of 5 $cm^3$ of culture plus 10 $cm^3$ of chromium trioxide (8% w/v). Heat the mixture to 70°C for 2 min in a preheated oven, cool it to room temperature, and shake the mixture vigorously on a bench-top shaker for approximately 10 min (Street, 1977). Place a droplet of the mixture on a microscope slide and examine it for evidence of cell separation. If cell aggregates are still present, return the 15-$cm^3$ mixture of cells and chromium trioxide to the 70°C oven for an additional 5–10 min. Another maceration reagent comprises a mixture of equal parts chromium trioxide (10% w/v) and HCl (10% v/v); it can be used in a 1:1 ratio with the culture. It is important that a homogeneous single cell sample be taken of the

suspension culture, and that the culture be *thoroughly* agitated at the time of the removal of the sample by pipette.

3. Various types of hemocytometer slides can be used, although nearly all of these devices have very small volumes for plant cell counting. Possibly the best aid is the Sedgwick–Rafter slide that was originally devised for counting aquatic microorganisms. This chamber (50 × 20 × 1 mm) has an area of 1,000 mm$^2$ and a total volume of 1.0 cm$^3$. Because the chamber has no grid markings, it is advisable to use a Whipple disk, which subdivides the optical field into squares. The Whipple disk is mounted inside the ocular of the microscope. In preparation for counting, it is necessary to determine the area covered in one optical field; that is, the field of vision seen under the Whipple disk at approximately 100× magnification (10× ocular and 10× objective). The diameter of the field is measured with a stage micrometer and the area is calculated. Next, decide on a system for selecting 10 positions in the chamber to be counted (e.g., you might count five different positions across the center of the chamber and another five positions along the periphery). It will be necessary to refocus the microscope at each position in order to count all the cells because the cells will be in different planes of focus. Obtain a mean value for the number of cells in the 10 Whipple fields.

4. The calculations are made as follows. The area of the chamber is 1,000 mm$^2$. If the area of the Whipple field is 1.86 mm$^2$, there are a total of 537.6 Whipple fields in the entire chamber (1,000/1.86). Let $N$ represent the mean cell count per Whipple field. The density $d$ of the culture, given in cells per cubic centimeter, will be

$$d = 537.6(N)$$

If the mean cell count per field is 15 cells, then the culture density is equal to 8,064 cells per cubic centimeter. If the sample has been diluted before counting, the density then equals 537.6($N$)(DF), where DF is the dilution factor. Dilution of the sample with water (1:1) gives DF = 2.

*Experiment 4. Determination of packed cell volume (PCV)*

1. Pipette from the suspension culture a known volume to be sample, and transfer it to a graduated conical centrifuge tube. More accurate results may be obtained with a tube consisting of an en-

larged upper part, the receiver, and a lower portion that is a calibrated capillary tube. These tubes have been employed by biologists to measure the PCV of algal suspension cultures (Sorokin, 1973). Depending on the density of the culture, it may be necessary either to dilute or concentrate the sample before placing it in the tube. For example, a relatively large sample (100 cm$^3$) can be given a precentrifugation to condense the cells in order for an appropriate reading to appear on the calibrated scale. Any dilutions or concentrations must be taken into consideration in the final calculation of the PCV.

2. Place the sample in the receiver of the centrifuge tube and spin it at 2,000 $g$ for a period of 5 min. If this does not provide a clear demarcation of the cell boundary, then increase the time of centrifugation. The readings must be made immediately following the centrifugation because any delay will give inflated values owing to cell expansion. Several replications should be run simultaneously and a mean value determined.

The results are expressed as cubic centimeters of cell pellet per cubic centimeter of culture.

## Experiment 5. Determination of mitotic index (MI)

Because of the difficulties encountered in the maceration of callus, it is advisable that the student perform this experiment on a sample of actively dividing suspension culture. The following procedure has been adapted from Wetter and Constabel (1982).

1. The carbol-fuchsin technique requires the preparation of three stock solutions. Stock A consists of basic fuchsin (3 g) dissolved in 100 cm$^3$ ethanol (70% v/v), and this solution is completely stable. Stock B is prepared by adding 10 cm$^3$ of stock A to 90 cm$^3$ phenol (5% w/v); this reagent should be discarded after two weeks. Stock C consists of 45 cm$^3$ of stock B, 6 cm$^3$ glacial acetic acid, and 6 cm$^3$ formaldehyde (37% v/v); this was employed as a carbol-fuchsin stain by Carr and Walker (1961). The present experiment uses a modified carbol-fuchsin stain prepared by adding 2–10 cm$^3$ of stock C to 90–98 cm$^3$ acetic acid (45% v/v) and 1.8 g sorbitol. The optimum concentration of the stain varies from 2% to 10%, depending on the plant species to be stained. The staining reagent is more effective if it is prepared at least two weeks in advance and stored at room temperature (Kao, 1982).

2. Pipette a 5-cm$^3$ sample from the culture and fix the cells in 1 cm$^3$ of a mixture of glacial acetic acid and ethanol (95% v/v) in a ratio of 1:3. Transfer a droplet of the fixed cells to a microscope slide and add an equal amount of the modified carbol-fuchsin stain. Wait about 5 min and then place a strip of lens paper on the stained preparation. This serves to remove the excess fluid from the surface of the slide. Add a cover slip atop the lens paper and squash the preparation by gently rolling a short glass rod over the cover slip. The red-stained nuclei will be visible in the microscope. The lens paper should not prove to be a hindrance since it will be in a slightly different plant of focus.

3. Examine microscopically a minimum of 500 nuclei and categorize them as either (a) nondividing, or (b) exhibiting some stage of mitosis (prophase–telophase). The student should be provided with a set of photomicrographs showing plant cell mitosis so that the early and late stages will not be overlooked.

The mitotic index (MI) represents the percentage of total nuclei displaying mitosis at a given time in a sample:

$$MI = \frac{\text{number of nuclei in mitosis}}{\text{total number of nuclei examined in sample}} \times 100$$

## Experiment 6. Tracheary element counts in primary explants

This experiment involves the maceration of primary explants and the associated callus following culture for varying periods of time on a xylogenic medium. The initial procedure should be followed as outlined in Chapter 12 on the induction of xylem differentiation.

1. Sacrifice the explants after 5, 10, and 15 days of incubation, and examine 10 explants for each time period. Place the explants individually in small vials and immerse them in 1 cm$^3$ of a maceration reagent consisting of chromium trioxide and hydrochloric acid (Steward & Shantz, 1956). They should remain in the maceration reagent for 24 hr at room temperature (Brown & Rickless, 1949). Because of the toxicity of the fumes, the maceration reagent should be added to the vials by Pasteur pipette in the hood. Stopper the vials, and, if possible, store them in the hood.

2. After the maceration period, withdraw the acid mixture and replace it with approximately 0.5 cm$^3$ distilled water. Care must be

exercised not to touch the fragile explants with the tip of the Pasteur pipette and not to remove any tissue that has crumbled from the sample. Attach a No. 22-gauge needle to a syringe and break up the sample with the tip of the needle until the disrupted cells can be drawn into the syringe. Expel the cells back into the vial. Repeat this pumping action until the sample has been thoroughly homogenized.

3. Transfer the macerated sample by syringe to a volumetric tube calibrated to 2 $cm^3$. Add distilled water until the level is precisely at the 2-$cm^3$ calibration mark. Add the cover slip diagonally across the Sedgwick–Rafter chamber so that spaces remain open at the opposite corners. *Note:* A suitable volumetric tube can be made from one of the small vials. Pipette 2 $cm^3$ of water into the vial. Mark the water level by attaching a strip of opaque plastic tape to the bottom of the meniscus.

4. Attach a No. 18-gauge needle to the syringe and thoroughly mix the sample by pumping it repeatedly in and out of the syringe. Deliver 1 $cm^3$ of the sample to the Sedgwick–Rafter chamber and slide the cover slip into the proper position (Guillard, 1973). If available, a green filter will aid in the detection of the tracheary elements.

5. The counting technique and calculations are similar to the procedure outlined in Experiment 3. The dilution factor, however, must be taken into account in the present calculations. The final volume of the sample was 2 $cm^3$, and DF = 2. The total number of tracheary elements in the sample ($\Sigma TE$) may be expressed as

$$\Sigma TE = 537.6(N)(DF)$$

If the mean tracheary element count per Whipple field is 20, then the sample contains 21,504 tracheary elements. If the sample has extremely large numbers of tracheary elements, then it is necessary to dilute the sample to a final volume of 4 or possibly 8 $cm^3$. For greater accuracy in counting, the number of tracheary elements in a Whipple field should not exceed roughly 25–30.

### QUESTIONS FOR DISCUSSION

1. What are some reasons why the fresh weight and dry weight of a callus sample might not be reliable measurements of the growth rate of the culture?

2. In regard to quantitation, what are some characteristics that algal cultures and higher plant cell cultures have in common?
3. In addition to the use of chromium trioxide, what are some other methods that have been used for the separation of plant cells?
4. What are some sources of experimental error involved in the use of the Sedgwick-Rafter chamber?
5. What information about a given culture does the mitotic index give the investigator? Why is the MI not a reliable indicator of the degree of cell division synchrony of a culture?

SELECTED REFERENCES

Brown, R., & Rickless, P. A. (1949). A new method for the study of cell division and cell extension with preliminary observations on the effect of temperature and nutrients. *Proc. Roy. Soc. B 136*, 110-25.

Carr, D. H., & Walker, J. E. (1961). Carbol fuchsin as a stain for human chromosomes. *Stain Technol. 36*, 233-6.

Guillard, R. R. L. (1973). Division rates. In *Handbook of phycological methods: Culture methods and growth measurements*, ed. J. R. Stein, pp. 289-311. Cambridge: Cambridge University Press.

Kao, K. N. (1982). Staining methods for protoplasts and cells. In *Plant tissue culture methods*, 2d Ed. L. R. Wetter & F. Constabel, pp. 67-71. Saskatoon: National Research Council of Canada, Prairie Regional Laboratory.

McAlice, B. J. (1971). Phytoplankton sampling with the Sedgwick-Rafter cell. *Limnol. Oceanogr. 16*, 19-28.

Phillips, R., & Dodds, J. H. (1977). Rapid differentiation of tracheary elements in cultured explants of Jerusalem artichoke. *Planta 135*, 207-12.

Sorokin, C. (1973). Dry weight, packed cell volume and optical density. In *Handbook of phycological methods: Culture methods and growth measurements*, ed. J. R. Stein, pp. 321-43. Cambridge: Cambridge University Press.

Steward, F. C., & Shantz, E. M. (1956). The chemical induction of growth in plant tissue cultures: Part 1. Methods of tissue culture and the analysis of growth. In *The chemistry and mode of action of plant growth substances*, ed. R. L. Wain & F. Wightman, pp. 165-86. London: Butterworths Scientific Publications.

Street, H. E. (1968). The induction of cell division in plant cell suspension cultures. In *Les cultures de tissus de plantes: First international conference on the culture of plant tissues* (Strasbourg, 1967), pp. 177-93. Strasbourg: Colloques Internationaux du CNRS.

(1977). Cell (suspension) cultures – Techniques. In *Plant tissue and cell cultures*, 2d Ed., ed. H. E. Street, pp. 61-102. Oxford: Blackwell Scientific Publications.

Wetter, L. R., & Constabel, F. (eds.) (1982). *Plant tissue culture methods*, 2d

rev. Ed. Saskatoon: National Research Council of Canada, Prairie Regional Laboratory.
White, P. R. (1943). *A handbook of plant tissue culture.* Tempe: Cattell Press.
(1963). *The cultivation of animal and plant cells,* 2d Ed. New York: Ronald Press.
Yeoman, M. M., & Evans, P. K. (1967). Growth and differentiation of plant tissue cultures: II. Synchronous cell divisions in developing callus cultures. *Ann Bot. (Lond.) 31,* 323–32.
Yeoman, M. M., Evans, P. K., & Naik, G. G. (1966). Changes in mitotic activity during early callus development. *Nature 209,* 1115–16.
Yeoman, M. M., & Macleod, A. J. (1977). Tissue (callus) cultures – Techniques. In *Plant tissue and cell culture,* 2d Ed., ed. H. E. Street, pp. 31–59. Oxford: Blackwell Scientific Publications.

# Abbreviations

| | |
|---|---|
| ABA | abscisic acid |
| AC | activated charcoal |
| B5 | Gamborg, Miller, & Ojima's (1968) medium |
| BA | benzyladenine |
| BAP | benzylaminopurine |
| $C_4$ | $C_4$ dicarboxylic acid pathway |
| CCC | (2-chloroethyl)-trimethylammonium chloride |
| cm | centimeter |
| 4-CPA | 4-chlorophenoxyacetic acid (also PCPA) |
| $d$ | density of culture (cells/cm$^3$) |
| 2,4-D | 2,4-dichlorophenoxyacetic acid |
| DDH$_2$O | double-distilled water |
| DF | dilution factor |
| DMSO | dimethyl sulfoxide |
| DNA | deoxyribonucleic acid |
| EDTA | ethylenediaminetetraacetic acid |
| $g$ | gravity |
| g | gram |
| gal | gallon |
| Gl | interval between mitosis (M) and DNA synthesis (S) in the cell cycle |
| GA$_3$ | gibberellic acid |
| GLC | gas–liquid chromatography |
| HEPA | high-efficiency particulate air |
| HPLC | high-performance liquid chromatography |
| hr | hour |
| IAA | indole-3-acetic acid |
| IAPTC | International Association for Plant Tissue Culture |
| IBA | indole-3-butyric acid |
| I.D. | inside diameter |
| in. | inch |
| 2iP | 6($\gamma,\gamma$–dimethylallylamino)purine |
| IPA | $N^6$-[$\Delta^2$-isopentyl]adenine |
| K | kinetin; $N^6$-furfuryladenine |
| kPa | kilopascal |

# Abbreviations

| | |
|---|---|
| lb | pound |
| $\mu$l | microliter |
| $\mu$W | microwatt |
| M | molar (concentration) |
| MES | 2-[N-morpholino]ethanesulfonic acid |
| mg | milligram |
| MI | mitotic index |
| mil | $10^{-3}$ inch |
| min | minute |
| ml | milliliter |
| mm | millimeter |
| MS | Murashige & Skoog's (1962) medium |
| N | mean cell count per Whipple field |
| N | normal (concentration) |
| NAA | naphthaleneacetic acid ($\alpha$- or $\beta$-) |
| nm | nanometer |
| O.D. | outside diameter |
| PABA | $p$-aminobenzoic acid; vitamin $B_x$ |
| PCPA | $p$-chlorophenoxyacetic acid (also 4-CPA) |
| PCV | packed cell volume |
| PEG | polyethylene glycol |
| RNA | ribonucleic acid |
| rpm | revolutions per minute |
| sec | second |
| S phase | DNA synthesis during cell cycle |
| STE | total number of tracheary elements in a given sample |
| 2,4,5-T | 2,4,5-trichlorophenoxyacetic acid |
| T-DNA | DNA from Ti plasma moved during transformation |
| TDZ | thidiazuron |
| TE | tracheary element(s) |
| UV | ultraviolet light |
| v/v | percent "volume in volume"; number of cubic centimeters of a constituent in 100 cm$^3$ of solution |
| W | watt |
| w/v | percent "weight in volume"; number of grams of constituent in 100 cm$^3$ of solution |

# Glossary

*adenine*  Aminopurine; exhibits cytokinin activity in bud initiation.

*adventitious*  Initiation of a structure out of its usual place, i.e., arising sporadically. Adventitious roots may originate from leaf or stem tissue.

*aseptic*  Sterile; free from contamination by microorganisms.

*auxin*  Plant growth regulator stimulating shoot cell elongation and resembling IAA in physiological activity.

*axenic*  Aseptic; free from other living organisms.

*batch culture*  Cell suspension grown in a fixed volume of liquid medium; example of a closed culture.

*boring platform*  Sterile bottom half of a Petri dish used for preparing tissue samples with a cork borer.

*callus*  Disorganized meristematic or tumorlike mass of plant cells formed in vitro; also refers to a meristematic growth arising at a wound site in vivo.

*cell cycle*  Sequence of events occurring during cell division, and measured by the time interval between one of these events and a similar event in the next generation. The demonstrable phases include mitosis (M) and DNA synthesis (S). The time period between S and M is termed $G_2$, whereas $G_1$ represents the interval between M and S.

*chemically defined medium*  A medium devoid of any natural plant or animal undefined organic supplement. In a strict sense, "a chemically defined medium" does not exist: All media contain traces of organic and inorganic contaminants.

*chemostat*  Instrument for maintaining an open, continuous culture; growth rate and cell density are maintained constant by regulating the input of a growth-limiting nutrient.

*chromic acid*  Aqueous solution of chromium trioxide.

*clones*  Genetically identical plants vegetatively propagated from a single individual.

*closed continuous culture*  Suspension culture in which the influx of fresh medium equals the efflux of spent medium; all cells are retained within the system.

*coconut water*  Liquid endosperm of coconut; coconut milk.

*continuous culture*  Suspension culture continuously supplied with an influx of fresh medium and maintained at a constant volume.

*cryopreservation*  Freeze preservation of plant material; typically maintained at the temperature of liquid nitrogen (–196 °C).
*cybrid*  Cytoplasmic hybrid; heteroplast.
*cytodifferentiation*  Cell differentiation; morphological and biochemical specialization of a cell; also the developmental process leading to this condition.
*cytokinin*  Plant growth regulator stimulating cell division and resembling kinetin in physiological activity; mainly $N^6$-substituted aminopurines.
*determination*  Resumption of reprogramming by which a cell becomes restricted to a new pathway of specialization.
*embryogenesis*  Initiation of embryo formation.
*embryoid*  Embryolike structure formed in vitro. With proper care, an embryoid will develop into a normal plant.
*explant*  Excised fragment of plant tissue or organ used to start a tissue culture; primary explant.
*fermenter*  Instrument for culturing a cell suspension under batch or continuous culture conditions.
*friable*  Crumbling or fragmenting easily.
*genetic engineering*  Use of a vector (e.g., plasmid) for the transfer of genetic information.
*gibberellin*  Plant growth regulator with physiological activity similar to gibberellic acid ($GA_3$).
*glycerol*  Glycerin; 1,2,3-propanetriol.
*glycine*  Aminoacetic acid.
*growth*  Process of accumulation of the cultured product; compare "yield."
*habituation*  Changes in exogenous nutritional requirements occurring during culture, particularly independent of external growth regulators.
*hairy root cultures*  Genetic transformation by *Agrobacterium rhizogenes* that results in cultured roots showing a high degree of branching, a profusion of root hairs, and a high rate of biosynthesis of secondary metabolites.
*haploid*  Having a single set of chromosomes; monoploid.
*hardening*  Mild environmental stress that prepares plants for more rigorous growing conditions; e.g., plantlets are given greater illumination, lower nutrient levels, and less moisture to enhance their survival outside of the culture tube.
*heterokaryon*  Fusion of unlike cells with dissimilar nuclei present; heterokaryocyte.
*heteroplast*  Cell containing foreign organelles; cytoplasmic hybrid.
*homokaryon*  Fusion of similar cells.
*induction*  Initiation of the formation of an organ or process in vitro, e.g., cell differentiation.
*inositol*  *myo*-Inositol; cyclohexitol.
*in vitro*  In an artificial environment outside the living organism.
*in vivo*  Within the living organism.
*meristem culture*  Apical meristem culture; explant consisting only of apical dome tissue distal to the youngest leaf primordium.
*meristemoid*  Cluster of meristematic cells within a callus with the potential to form a primordium.
*micropropagation*  Clonal multiplication of plants originating from cultured shoot tips. Cytokinin-induced proliferation of shoots occurs during subcultures.
*mutagen*  Chemical or physical treatment for the induction of gene mutation.

*nicotinic acid*  Niacin; pyridine-β-carboxylic acid.
*open continuous culture*  Suspension culture in which the influx of fresh liquid medium is equal to the efflux of suspension culture and in which the culture (i.e., the cells) are flushed out with the spent medium.
*osmoticum*  Isotonic plasmolyticum; external medium of low osmotic potential approximating the concentration of solutes within the cell, thereby preventing bursting of protoplasts due to excessive water uptake.
*passage*  Subculture; transfer of inoculum from a culture to a fresh medium.
*passage number*  Number of subcultures completed.
*passage time*  Interval between successive subcultures.
*pectinase*  Polygalacturonase; enzyme liberating galacturonic acid from polygalacturonic acid (E.C. 3.2.1.15).
*photoautotropic culture*  Suspension culture in which an exogenous carbon source is not required in the medium. $CO_2$ fixation by photosynthesis provides sufficient carbohydrates.
*picloram*  4-amino-3,5,6-trichloropicolinic acid.
*plantlet*  Miniature plant with root and shoot systems regenerated by tissue culture.
*primordium* (pl. *primordia*)  Earliest detectable stage of development of an organ, usually pertaining to leaf and bud primordia.
*protoplast*  Living isolated plant cell following removal of cell wall either by enzymatic or mechanical method.
*pyridoxine*  Vitamin $B_6$.
*rhizogenesis*  Initiation of adventitious root primordia.
*secondary metabolite*  Metabolic compound unique to a limited number of species or cultivars.
*sector inoculum*  Fragment of main root and lateral roots used to initiate a root subculture.
*shoot-apex culture*  Explant consisting of the apical dome plus a few subjacent leaf primordia.
*somatic embryogenesis*  Formation of embryos in vitro from vegetative (asexual) cells.
*somatic hybrid*  Cell or plant derived from the fusion of two genetically different vegetative (somatic) protoplasts.
*subculture*  See "passage"; transfer of inoculum to a fresh medium.
*suberization*  Conversion of cell walls into corky tissue via infiltration with suberin.
*synchronous culture*  Population of individual cells brought into phase or synchrony, i.e., passing through the sequential events of the cell cycle at the same time.
*synkaryocyte*  Hybrid cell produced by fusion of nuclei in a heterokaryon.
*thiamine*  Vitamin $B_1$.
*totipotency*  Retention by a single cell of the genetic information for the recreation of the adult organism.
*transdifferentiation*  Differentiation of a cell into a new cell type without undergoing cell division.
*transformation*  Insertion of foreign DNA into a plant cell resulting in a phenotypic modification of the regenerated plant.
*turbidostat*  Instrument for growing an open continuous culture into which fresh medium flows due to changes in culture turbidity, i.e., cell density.

*Type I callus*   Type of adventive embryogenesis in gramineous monocots characterized by somatic embryos arrested at the coleoptilar or scutellar stage of development. Embryos may be fused, particularly by a common coleorhiza. Subcultured inocula maintain this trait.

*Type II callus*   Type of adventive embryogenesis in gramineous monocots characterized by somatic embryos arrested at the globular stage. Globular embryos often arise separately from a common base. Subcultured inocula maintain this morphology.

*yield*   Net result of the metabolic processes of a culture after a given time period; compare "growth."

# Commercial sources of supplies

Within the text references are made to special chemicals, supplies, and equipment. Here is a brief list of the names and addresses of some of the suppliers. The inclusion of a particular supplier does not imply endorsement by the authors exclusive of other companies offering a similar product.

### *Abrasives*

Norton Consumer Products, One New Bond Street, Worcester, MA 01606

### *Agar, gels, and premixed media*

Carolina Biological Supply Company, 2700 York Road, Burlington, NC 27215, USA
Difco Laboratories Inc., P.O. Box 331058, Detroit, MI 48232-7058, USA
Difco Laboratories Ltd., P.O. Box 14B, Central Avenue, East Molesey, Surrey KT8 0SE, UK
FMC BioProducts, 191 Thomaston Street, Rockland, ME 04841, USA
FMC BioProducts Europe, 1 Risingevej, DK-2665, Vallensbaek Strand, Denmark
GIBCO BRL, Life Technologies, 3175 Staley Road, P.O. Box 68, Grand Island, NY 14072-0068, USA
GIBCO BRL, Life Technologies Ltd., P.O. Box 35, Trident House, Renfrew Road, Paisley PA3 4EF, Scotland
Miles Laboratories, Elkhart, IN 46514, USA
Miles Ltd., Stock Court, Stoke Poges, Slough, Berkshire SL2 4LY, UK
Oxoid USA Inc., 9017 Red Branch Road, Columbia, MD 21045, USA
Oxoid Ltd., Wade Road, Basingstoke, Hants, RG24 0PW, UK
Sigma Chemical Company, P.O. Box 14508, St. Louis, MO 63178-9916, USA
Sigma Chemical Company Ltd., Fancy Road, Poole, Dorset BH17 7NH, UK

### *Chemicals*

Aldrich Chemical Company Inc., 1001 West St. Paul Avenue, P.O. Box 355, Milwaukee, WI 53201, USA

J. T. Baker Inc., 222 Red School Lane, Phillipsburg, NJ 08865, USA
Calbiochem, P.O. Box 12087, San Diego, CA 92112-9939, USA
Fisher Scientific, 711 Forbes Avenue, Pittsburgh, PA 15219, USA
Gallard-Schlesinger Industries Inc., 584 Mineola Avenue, Carle Place, NY 11514, USA
ICN Biomedicals Inc., Marketing Department, P.O. Box 19536, Irvine, CA 92713-9921, USA
Mallinckrodt Inc., Science Products Division, 675 McDonnell Blvd., P.O. Box 5840, St. Louis, MO 63134, USA
Sigma Chemical Company, P.O. Box 14508, St. Louis, MO 63178-9916, USA
Sigma Chemical Company Ltd., Fancy Road, Poole, Dorset BH17 7NH, UK
United States Biochemical Corp., P.O. Box 22400, Cleveland, OH 44122, USA

## Cellulose cell wall stain

Polysciences Inc., 400 Valley Road, Warrington, PA 18976, USA (Calcofluor White M2R)
Sigma Chemical Company, P.O. Box 14508, St. Louis, MO 63178-9916, USA (Fluorescent brightener)

## Cell wall digestive enzymes

Calbiochem, P.O. Box 12087, San Diego, CA 92112-9939, USA (Cellulysin; Macerase)
Sigma Chemical Company, P.O. Box 14508, St. Louis, MO 63178-9916, USA (Cellulysin, Driselase, Hemicellulase, Macerase, Pectinase, Pectolyase)

## Cultures

American Type Culture Collection [ATCC], 12301 Parklawn Drive, Rockville, MD, 20852, USA

## Filters and screens

Gelman Sciences Inc., 600 S. Wagner Road, Ann Arbor, MI 48106, USA
Micron Separations Inc., P.O. Box 1046, 135 Flanders Road, Westboro, MA 01581, USA
Millipore Corporation, 80 Ashby Road, P.O. Box 9125, Bedford, MA 01730-9125, USA
Millipore (UK) Ltd., 11-15 Peterborough Road, Harrow, Middlesex HA1 2YH, UK
Schleicher & Schuell, 10 Optical Avenue, P.O. Box 2021, Keene, NH 03431, USA
Schleicher & Schuell GmbH, P.O. Box 4, D-3354, Dassel, Germany
Tetko Inc., Precision Woven Screening Media, P.O. Box 183, 111 Calumet Street, Lancaster, NY 14086, USA
Whatman Speciality Products Inc., 6 Just Road, Fairfield, NJ 07004, USA
Whatman Ltd., Springfield Mill, Maidstone, Kent ME14 2LE, UK

## Germicidal lighting

Vangard International, Inc., 111-A Green Grove Road, P.O. Box 308, Neptune, NJ 07754-0308, USA
Westinghouse Electric Corp., Lamp Commercial Division, Bloomfield, NJ 07003, USA

## Glassware and plastics

Bellco Glass Inc., P.O. Box B, 340 Edrudo Road, Vineland, NJ 08360, USA
Cole-Parmer Instrument Company, 7425 North Oak Park Avenue, Chicago, IL 60648, USA
Corning Inc., Science Products, MP-21-5-8, Corning, NY 14831, USA
Costar Corp., One Alewife Center, Cambridge, MA 02140, USA
Kontes, Spruce Street, P.O. Box 729, Vineland, NJ 08360, USA
Nalge Company, A Subsidiary of Sybron Corp., P.O. Box 20365, Rochester, NY 14602-0365, USA
Vangard International, 1111-A Green Grove Road, P.O. Box 308, Neptune, NJ 07754-0308, USA
Wheaton, A Division of Wheaton Industries, 1301 N. Tenth Street, Millville, NJ 08332, USA

## Laminar flow hoods and glove boxes

The Baker Company, P.O. Drawer E., Sanford, ME 04073, USA
Bellco Glass Inc., P.O. Box B, 340 Edrudo Road, Vineland, NJ 08360, USA
Carolina Biological Supply Company, 3700 York Road, Burlington, NC 27215, USA
Clean Room Products Inc., 1800 Ocean Avenue, Ronkonkoma, NY 11779-6528, USA
Forma Scientific Inc., Millcreek Road, Box 649, Marietta, OH 45750, USA
Germfree Laboratories, 7435 NW 41 Street, Miami, FL 33166, USA
Labconco Corporation, 8811 Prospect Avenue, Kansas City, MO 64132, USA
National Labnet Company, P.O. Box 841, Woodbridge, NJ 07095, USA

## Light microscopes

Leica Inc., 111 Deer Lake Road, Deerfield, IL 60015, USA (includes Cambridge Instruments, Jung, Kern, Leitz, Reichert, and Wild Heerbrugg)
Nikon Inc., 1300 Walt Whitman Road, Melville, NY 11747-3064, USA
Nikon Europe B.V., Instrument Department, Schipholweg 321, P.O. Box 222, NL-1170 AE Badhoevedorp, The Netherlands
Olympus Corporation, Precision Instruments Division, 4 Nevada Drive, Lake Success, NY 11042-1179, USA
Carl Zeiss Inc., Microscope Division, One Zeiss Drive, Thornwood, NY 10594, USA

## Magnifiers with illuminators

Edmund Scientific Company, 101 E. Gloucester Pike, Barrington, NJ 08007-1380, USA

Manning Holoff Company, 14603 Arminta Street, Van Nuys, CA 91402, USA

## Nursery supplies

Hyponex Corp., 14111 Scottslawn Road, Marysville, OH 43041

L&L Nursery Supply, Inc., 5354 G Street, Chino, CA 91710

## Parafilm

Baxter Export Corporation, 8350 NW 52nd Terrace, Suite 300, Miami, FL 33166, USA

## Polypropylene film

Norprop Films, 601 East Lake Street, Streamwood, IL 60103, USA (Norprop 380CW, 0.80 mil polypropylene film)

## Seeds

W. Atlee Burpee & Co., Commercial Sales Department, 300 Park Avenue, Warminster, PA 18991-0002, USA

Earl May Seed & Nursery, 208 N. Elm, Shenandoah, IA 51603, USA

Gurney Seed Company, 110 Capitol Street, Yankton, SD 57078, USA

Thompson & Morgan Inc., P.O. Box 100, Farmingdale, NJ 07727, USA

## Surgical instruments

Becton Dickinson & Co., Clay Adams Division, 299 Webro Road, Parsippany, NJ 07054, USA

Carolina Biological Supply Company, 2700 York Road, Burlington, NC 27215, USA

Techi-Tool, 5 Apollo Road, P.O. Box 368, Plymouth Meeting, PA 19462, USA

## Tissue culture apparatus

Bellco Glass Inc., P.O. Box B, 340 Edrudo Road, Vineland, NJ 08360, USA

Carolina Biological Supply Company, 2700 York Road, Burlington, NC 27215, USA

Corning Inc., Science Products, MP-21-5-8, Corning, NY 14831, USA

Costar Corp., One Alewife Center, Cambridge, MA 02140, USA

Fisher Scientific, Inquiry Department, 711 Forbes Avenue, Pittsburgh, PA 15219-9919, USA

Germfree Laboratories, 7435 NW 41 Street, Miami, FL 33166, USA
New Brunswick Scientific Company Inc., 44 Talmadge Road, P.O. Box 4005, Edison, NJ 08818, USA
Wheaton, A Division of Wheaton Industries, 1301 N. Tenth Street, Millville, NJ 08332, USA

## Water purification systems

Barnstead-Thermolyne Corporation, 2555 Kerper Blvd., P.O. Box 797, Dubuque, IA 52004-0797, USA
Bellco Glass Inc., P.O. Box B, 340 Edrudo Road, Vineland, NJ 08360, USA
Corning Inc., Science Products, MP-21-5-8, Corning, NY 14831, USA
IonPure Technologies Corp., 10 Technology Drive, Lowell, MA 01851, USA

# Formulations of tissue culture media

| Ingredients | Formulations (mg/l) | | | |
|---|---|---|---|---|
| | MS[a] | B5[b] | WH[c] | SH[d] |
| $(NH_4)NO_3$ | 1,650 | | | |
| $(NH_4)_2SO_4$ | | 134 | | |
| $(NH_4)H_2PO_4$ | | | | 300 |
| $KNO_3$ | 1,900 | 2,528[e] | 80 | 2,500 |
| $Ca(NO_3)_2 \cdot 4H_2O$ | | | 300 | |
| $CaCl_2 \cdot 2H_2O$ | 440 | 150 | | 200 |
| $MgSO_4 \cdot 7H_2O$ | 370 | 246[e] | 720[e] | 400 |
| $Na_2SO_4$ | | | 200[e] | |
| $KH_2PO_4$ | 170 | | | |
| $NaH_2PO_4 \cdot H_2O$ | | 150 | 19 | |
| KCl | | | 65 | |
| $FeSO_4 \cdot 7H_2O$ | 27.8 | 27.8 | 27.8[f] | 15 |
| $Na_2EDTA \cdot 2H_2O$ | 37.2[e] | 37.2[e] | 37.2[f] | 20 |
| $MnSO_4 \cdot 4H_2O$ | 22.3 | | 7.0 | |
| $MnSO_4 \cdot H_2O$ | | 10 | | 10 |
| $ZnSO_4 \cdot 7H_2O$ | 8.6 | 2.0 | 3.0 | 1.0 |
| $H_3BO_3$ | 6.2 | 3.0 | 1.3 | 5.0 |
| KI | 0.83 | 0.75 | 0.75 | 1.0 |
| $MoO_3$ | | | 0.0001 | |
| $Na_2MoO_4 \cdot 2H_2O$ | 0.25 | 0.25 | | 0.1 |
| $CuSO_4 \cdot 5H_2O$ | 0.025 | 0.025 | 0.001 | 0.2 |
| $CoCl_2 \cdot 6H_2O$ | 0.025 | 0.025 | | 0.1 |
| *myo*-inositol | 100 | 100 | | 1,000 |
| nicotinic acid | 0.5 | 1.0 | 0.5 | 5.0 |
| pyridoxine·HCl | 0.5 | 1.0 | 0.1 | 0.5 |
| thiamine·HCl | 0.1 | 10.0 | 0.1 | 5.0 |
| glycine | 2.0 | | 3.0 | |
| sucrose | 30,000 | 20,000 | 20,000 | 30,000 |
| kinetin | 0.4-10 | 0.1 | | |
| IAA | 1-30 | | | |
| 2,4-D | | 0.1-1.0 | | 0.5 |
| 4-CPA | | | | 2.0 |
| pH | 5.75 | 5.5 | 5.5 | 5.9 |

*Notes to the Table*

[a]Murashige, T., & Skoog, F. (1962). A revised medium for rapid growth and bioassays with tobacco tissue cultures. *Physiol. Plant. 15,* 473–97.

[b]Gamborg, O. L., Miller, R. A., & Ojima, K. (1968). Nutrient requirement of suspension cultures of soybean root cells. *Exp. Cell Res. 50,* 151–8.

[c]White, P. R. (1963). *The cultivation of animal and plant cells,* 2d Ed. New York: Ronald Press. The present formulation is based on the corrections given by Owen, H. R., & Miller, A. R. (1992). An examination and correction of plant tissue culture basal medium formulations. *Plant Cell, Tissue & Organ Culture 28,* 147–50.

[d]Schenk, R. V., & Hildebrandt, A. C. (1972). Medium and techniques for induction and growth of monocotyledonous and dicotyledonous plant cell cultures. *Can. J. Bot. 50,* 199–204.

[e]See Owen & Miller's (1992) publication for details.

[f]White's (1963) medium required 2.5 mg $Fe_2(SO_4)_3$. Since ferric sulfate precipitates easily, it is more convenient to use the MS formulation in the chelated form.

# Author index

Aartrijk, J. van, 85
Abbott, A. J., 119
Abo El-Nil, M. M., 169
Ahkong, Q. F., 185
Aitchison, P. A., 70, 205
Akyar, O. C., 46
Akutsu, M., 206
Al-Abta, S., 104
Alfermann, A. W., 70, 206, 207
Almehdi, A. A., 60
Aloni, R., 156
Ammirato, P. V., 101, 104
Anagnostakis, S. L., 138
Anderson, W. C., 44
Ardenne, R. V., 207
Ausubel, F. M., 138
Aviv, D., 167, 187, 209

Baba, S., 49
Bachelard, E. P., 118
Backs-Hüsemann, D., 105
Bajaj, Y. P. S., 105, 137, 138, 171, 176, 195
Ball, E., 9, 10, 45, 53, 126
Bandurski, R. S., 46
Banks, M. S., 104
Barton, K. A., 12
Bastiaens, L., 35
Bayliss, M. W., 102
Beasley, C. A., 37
Behagel, H. A., 33
Bell, E. A., 204
Bengochea, T., 12, 168, 199
Bergmann, L., 8
Berman, R., 50
Berquam, D. L., 3, 4
Bhatnagar, S. P., 136
Bhojwani, S. S., 136, 169
Bilkey, P. C., 85
Binding, H., 178, 179
Binns, A. S., 48
Biondi, S., 21, 24, 27, 30
Blakeslee, A. F., 6
Bloksberg, L. N., 60

Blom-Barnhoorn, G. J., 85
Böhm, H., 207
Boll, W. G., 123
Bolwell, G. P., 156
Bolton, W. E., 33, 34
Bonga, J. M., 10, 24, 30, 44, 50, 51, 54
Bonner, J., 118
Bordon, J., 138
Boss, W. F., 46
Bottcher, U. F., 187
Boulay, M., 130
Bourgin, J. P., 9
Bragt, J. van, 52
Braun, A. C., 6, 10, 69
Bravo, J. E., 170, 178
Bressan, R. A., 73
Bridgen, M. P., 198
Brill, W. J., 12
Broad, W. J., 36
Brown, C. L., 118
Brown, D. C. W., 50
Brown, R., 217, 225
Brown, R. M., Jr., 158
Brown, S., 54
Burdon, J., 50
Burger, D. W., 29
Burgess, J., 158
Burk, L. G., 140
Burris, R. H., 45
Butcher, D. N., 10, 93, 118, 119, 205, 206, 207
Butenko, R. G., 11, 24, 34
Button, J., 102

Cahn, R. D., 32
Campbell, L. E., 23
Camus, G., 4
Caplin, S. M., 6, 53
Carlson, P. S., 168, 186
Carr, D. H., 218, 224
Cathey, H. M., 23
Chaleff, R. S., 11
Chandler, S. F., 206
Chandra, N., 102

Chaplin, J. F., 140
Chapman, H. W., 73
Cheng, T. Y., 52
Chilton, M. D., 12
Chlyah, A., 83, 84
Chlyah, H., 83
Chu, C-C., 44
Church, D. L., 158, 159, 160, 163
Clauss, H., 207
Cocking, E. C., 9, 136, 167, 168, 169, 170, 171, 173, 175, 179, 180, 185, 186
Collin, H. A., 104, 204
Collins, C. H., 26, 27, 30
Collins, G. B., 48, 139, 144
Combatti, N. C., 186
Conger, B. V., 10, 24, 105
Conklin, M. E., 6
Conover, R. A., 102
Constabel, F., 102, 170, 177, 178, 179, 183, 185, 215, 217, 224
Constantin, M. J., 10, 50
Cooke, R. C., 85
Coutts, R. H. A., 185
Cress, D. E., 168
Cummings, B. G., 85, 88
Cummins, S. E., 185
Cutter, E. G., 127

Dalessandro, G., 163
Dalton, C. C., 118
Darlington, D. C., 139, 144
Davey, M. R., 77, 85, 94, 167, 185
Davey, R., 139, 158
Dearing, R. D., 186
Debergh, P. C., 127, 128, 129
de Fossard, R. A., 21, 24, 34
Dekhuijzen, H. M., 53
Delmer, D. P., 56
De Ropp, R. S., 8
Devreux, M., 140, 141
Digby, J., 45
Dixon, R. A., 26, 54, 85
Dodds, J. H., 12, 35, 52, 126, 130, 146, 168, 169, 195, 197, 199, 206, 217
Doley, W. P., 49
Döller, G. von, 206
Doornbos, T., 53
Dougall, D. K., 54, 104
Dravnicks, D. E., 45
Drew, R. L. K., 104
Dudits, D., 167
Dunlap, J. R., 52
Dunn, D. M., 102
Dunwell, J. M., 137, 138, 145, 146
Durand-Cresswell, R., 130
Durzan, D. J., 10

Eaks, I. D., 37
Edwards, G. E., 169, 175
Eichholtz, D. A., 34
Einset, J. W., 49

El-Shagi, H., 206
Emmerich, A., 46
Eriksson, T., 45, 104, 185
Esau, K., 69, 157
Evans, D. A., 82, 170, 178
Evans, P. K., 167, 168, 170, 171, 175, 179, 218
Everett, N., 85

Falkiner, F. R., 34, 35, 37
Falconer, M. M., 158
Favero, M. S., 51
Feirer, R. P., 106
Fitter, M. S., 178
Fleurat-Lessard, P., 56
Flick, C. E., 82, 85
Foroughi-Wehr, B., 138
Fosket, D. E., 157
Fowler, M. H., 105
Francelet, A., 130
Fridborg, G., 50, 104
Fujimura, T., 104, 105
Fukuda, H., 155, 157, 158, 159, 160, 161, 163

Galletta, P. D., 31
Galston, A. W., 158, 160, 163
Galun, E., 167, 169, 187, 209
Gamborg, O. L., 44, 45, 48, 54, 102, 241
Gautheret, R. J., 4, 5, 6, 10, 11, 23, 34, 71, 78
Gleba, Y. Y., 175
Glimelius, K., 185
Goforth, P. L., 118
Goodwin, P. B., 128
Grasham, J. L., 9
Gresshoff, P. M., 84
Grout, B. W. W., 168, 185
Guha, S., 9, 136, 137
Guillard, R. R. L., 226
Guinn, G., 118
Gwynn, G. R., 140

Haberlandt, G., 4, 5
Hagen, S. R., 30, 33, 53, 73, 77
Haigler, C. H., 158
Halperin, W., 102
Hamilton, R. D., 26, 28, 30, 35, 36, 37
Hangarter, R. P., 44, 47, 53, 54
Hanson, M. R., 138
Hanus, F. J., 43
Harada, H., 85
Harney, P. M., 85, 88
Harvey, A. E., 9
Hasegawa, P. M., 34, 73
Hashimoto, T., 120
Heberle, E., 138
Helgeson, J. P., 10, 93
Heller, R., 52
Henke, R. R., 50
Henshaw, G. G., 50, 102, 138, 178, 197, 198, 206
Herr, J. M., 175
Hervey, A., 51

Hildebrandt, A. C., 7, 8, 54, 85, 169, 241
Hiraoka, N., 206
Ho, W. Jane, 102
Hoffmann, F., 138
Holder, A. A., 12
Homès, J., 102
Hopgood, M. E., 196
Horner, M., 139
Horsch, R. B., 52
Hu, C. Y., 129, 130
Hughes, K. W., 85
Hu Han, H. T. Y., 139
Huang, L. C., 45, 49, 50
Huntrieser, I., 84
Hussey, G., 128
Huxter, T. J., 85

Ikeda, M., 44
Ingold, E., 158
Ingram, D. S., 10, 93
Igbal, K., 118
Iwasaki, T., 156

Jablonski, J., 4, 5
Jacobsen, E., 138
Jacobson, E. D., 37
Jeffs, R. A., 8
Jensen, C. J., 140
Jhang, J. J., 206
Jones, D., 31
Jones, G. E., 52
Jones, L. E., 8
Jones, O. P., 196

Kamada, H., 85
Kameya, T., 176
Kamimura, S., 206
Kanai, R., 169, 175
Kao, K. N., 179, 185, 186, 218, 224
Kasperbauer, M. J., 137, 139, 144
Kaul, B., 206
Keller, W. A., 141, 185
Kim, J. Y., 206
King, J., 52
King, P. J., 92, 93
Klein, D. T., 26, 49, 55
Klein, R. M., 26, 49, 55, 205
Klucas, R. V., 43
Knap, A., 85, 88
Kobayashi, H., 158
Kochba, J., 102
Kohlenbach, H. W., 50, 101, 102, 103, 138, 143, 155, 157
Komamine, A., 104, 105, 155, 157, 158, 159, 161, 163
Komor, E., 45
Konar, R. N., 102
Kordan, H. A., 30, 71
Koritsas, V. M., 156
Kotte, W., 5
Krell, K., 37
Kresovich, S., 52

Krikorian, A. D., 3, 4, 7, 178
Kuster, E., 183

Labib, G., 9, 185, 187
LaCour, L. F., 139, 144
Laneri, V., 141
Langhans, R. W., 129
Larkin, P. J., 175
LaRosa, P. C., 73
Lay, J., 158
Leifert, C., 30, 34
Leung, D. W. M., 50
Levitt, J., 50
Lin, M. L., 208
Linsmaier-Bednar, S. M., 45
Linstead, P., 158
Litvay, J. D., 106
Litz, R. E., 102
Lloyd, G., 44
Loewus, F. A., 46
Loo, S., 11
Loo, S. W., 126
Loomis, R. S., 118
Lörz, H., 176
Lu, D. Y., 178
Luckner, M., 204
Lurquin, P. F., 168
Lutz, J., 48
Lyne, P. M., 26, 27, 30

McAlice, B. J., 217
McCown, B., 44
McCown, B. H., 85
McCully, M. C., 78
McFadden, J. J., 45
McGee, R. E., 52
Macleod, A. J., 69, 70, 71, 205, 216, 218
McWilliam, A. A., 102
Maene, L. J., 127, 129
Maheshwari, S. C., 9, 136, 137
Mahlberg, P. G., 38, 39
Maneval, W. E., 5
Mansur, M. A., 50
Mapes, M. O., 98, 101
Mark, J. B., Jr., 50
Marschner, H., 43
Martin, S. M., 59, 205
Martineau, B., 138
Masi, P., 38
Mastrangelo, I. A., 183
Mattson, J. S., 50
Mazur, P., 99
Mears, K., 101
Meins, F., Jr., 48
Melchers, G., 9, 12, 185, 187
Merkel, N., 8
Michayluk, M. R., 179, 185
Mignon, G., 106
Miller, A. R., 44, 47, 49, 57, 59, 156, 241
Miller, C. O., 7, 83
Miller, L. R., 54

Miller, R. A., 44, 102
Misawa, M., 12, 206
Mix, G., 138
Mizuguchi, R., 70
Moore, M. B., 102
Morel, G., 8, 10, 46, 126, 127, 128, 130
Morgan, T. H., 3
Morré, D. J., 46
Mothes, K., 204
Muir, W. H., 4, 7, 8
Muller, J. R., 46
Murashige, T., 11, 22, 23, 44, 45, 49, 54, 56, 57, 117, 126, 127, 128, 129, 131, 132, 241
Mutaftschiev, S., 84

Nagata, T., 175, 185
Nagmani, R., 144
Naik, G. G., 218
Nakagawa, H., 102
Nakata, K., 140
Narayanaswamy, S., 70, 101
Nataraja, K., 102
Nickell, L. G., 7, 205
Nishi, A., 97
Nishikawa, M., 206
Nissen, S. J., 47, 52, 53
Nitsch, C., 9, 44, 50, 136, 137
Nitsch, J. P., 44, 50
Nobécourt, P., 5, 6
Northcote, D. H., 8, 53, 158, 170
Nover, L., 204

O'Brien, T. P., 78
O'Hara, J. F., 178
Ohira, K., 44, 95
Ojima, K., 44, 95, 241
Owen, H. R., 44, 57, 59, 60, 241

Pareek, L. K., 102
Parfitt, D. E., 60
Patnaik, G., 170, 186
Paul, D. R., 38
Peer, H. G., 53
Pelet, F., 69
Phillips, G. C., 48
Phillips, R., 52, 217
Pierik, R. L. M., 44, 51, 52, 54, 129, 130
Pippard, D. J., 175
Poli, F., 56
Pontecorvo, G., 185
Pontikis, C. A., 196
Poovaiah, B. W., 45
Potrykus, I., 102, 176
Power, J. B., 167, 169, 173, 175, 185, 186, 187
Proskauer, K., 50
Puhan, Z., 205

Raghavan, V., 101, 138, 139, 144, 145, 146
Raj, B., 175
Rechinger, C., 4, 89
Reddy, A. S. N., 45

Redenbaugh, K., 186
Rédei, G. P., 53
Reid, D. M., 85
Reinert, J., 7, 8, 101, 102, 105, 107, 137, 138, 195, 207
Reinhard, E., 70, 206, 207
Reynolds, T. L., 146
Rhodes, M. J. C., 120
Rickless, P. A., 217, 225
Rier, J. P., 8
Riker, A. J., 7
Riopel, J. L., 55
Robbins, W. J., 5, 51
Roberts, L. W., 8, 35, 47, 49, 71, 80, 155, 156, 157, 163
Roberts, M., 138, 142, 143, 144, 145
Robitaille, H. A., 34
Roblin, G., 56
Rottier, J. M., 12
Russell, S. A., 43

Sacristan, M. D., 12
Said, A. G. E., 117
Saitou, T., 85
Sangwan-Noreel, B. S., 138
Saunders, J. W., 49
Schenk, R. U., 54, 241
Scherberger, R. F., 36
Schmidt, B., 155, 157
Schmitz, R. Y., 56
Schneider, V. F., 46
Schöpke, C., 155
Schulte, U., 206
Schwann, Th., 3
Scowcroft, W. R., 197
Seagull, R. W., 158
Seeliger, T., 118
Selby, K., 37
Shabde-Moses, M., 131, 132
Shantz, E. M., 225
Sharp, W. R., 82, 101, 102
Shaw, R., 73
Sheehy, R. E., 168
Shepard, J. F., 11, 169
Shibaoka, H., 156, 158
Shigomura, J., 8
Short, K. C., 85
Shyluk, J. P., 45
Sinnott, E. W., 69
Sinoto, Y., 102
Skirvin, R. M., 59, 60
Skoog, F., 5, 7, 44, 45, 50 56, 57, 82, 83, 241
Smith, H. H., 186
Smith, J., 98
Smith, R. H., 56, 126, 127, 131
Smith, S. M., 102, 105
Snow, R., 6
Sommer, H. E., 118
Sondahl, M. R., 102
Sopory, S. K., 138

Sorokin, C., 214, 216, 217, 224
Sorokin, S., 5, 8
Spiegel-Roy, P., 169
Staba, E. J., 205, 206, 208
Staby, G. L., 198
Stamp, J. A., 102
Start, N. D., 85, 88
Stasinopoulos, T. C., 44, 47, 53, 54
Steck, W., 206
Stehsel, M. L., 53
Steiner, A. A., 44
Sterling, C., 7
Steward, F. C., 6, 7, 98, 101, 102, 225
Stiff, C. M., 34
Stohs, S. J., 206
Stowe, B. B., 118
Strange, R. N., 175
Straus, J., 50
Street, H. E., 5, 7, 8, 19, 21, 24, 46, 48, 49, 51, 69, 71, 84, 93, 94, 98, 102, 105, 117, 118, 119, 139, 207, 216, 217, 218, 222
Strickland, R. G., 207
Stringham, G. R., 141
Strobel, G. A., 175
Sugano, N., 97
Sugiyama, M., 155, 157, 158, 159, 161
Sunderland, N., 136, 137, 138, 142, 143, 144, 145, 146, 207
Sutter, E., 129
Sutter, E. G., 47, 52, 53
Sweet, H. C., 33, 34
Sykes, G., 36

Tabata, M., 206
Tabor, C. A., 52
Takebe, I., 9, 175
Tanaka, M., 140
Thelen, M. P., 158
Thimann, K. V., 6
Thomas, E., 77, 94, 102, 126, 138, 139, 176
Thorpe, T. A., 10, 21, 24, 27, 30, 46, 50, 84, 85
Ten Ham, E. J., 53
Tisserat, B., 31, 33
Tomes, D. T., 10
Torres, K. C., 25, 32, 129
Torrey, J. G., 8, 118, 120
Totten, R. E., 169
Towill, L. E., 99
Tran Thanh Van, K., 83, 84
Tran Thanh Van, M., 84
Trécul, M., 4
Trinh, H., 84
Tsui, C., 7, 82
Tulecke, W., 9, 136
Turner, D. A., 118

Uchimiya, H., 176
Upadhya, M. D., 169

Van der Plas, L. H. W., 73
Vannini, G. L., 56
Van Overbeek, J., 6
van Winden, H., 44
Vardi, E., 169
Vasil, I. K., 10, 102, 136, 175
Vasil, V., 8, 175
Vasquez, A. M., 85
Veith, R., 45
Vian, W. E., 48
Vöchting, H., 4
Voqui, T. H., 52

Wade, N., 27
Waites, W. M., 30, 34
Wagner, M. J., 73
Walker, J. E., 218, 224
Wallin, S., 185
Wang, P. J., 50, 129, 130
Wardlaw, C. W., 9, 10
Warren, G. S., 105
Warren Wilson, J., 46
Watts, M., 204
Weatherhead, M. A., 50, 138
Weiler, E. W., 206
Welander, M., 84
Wengerd, D., 59
Went, F. W., 6
Wenzel, G., 138
Wernicke, W., 50, 138, 143
Westcott, R. J., 198, 206
Wetherell, D. F., 22, 24, 25, 54, 102, 104
Wetmore, R. H., 5, 8, 9, 10, 126
Wetter, L. R., 215, 217, 224
White, P. R., 3, 5, 6, 10, 11, 19, 23, 25, 42, 44, 118, 122, 126, 215, 241
Wicks, F. M., 143
Wilkins, C. P., 126, 130, 195, 197, 199
Willett, H. P., 28
Willison, J. H. M., 168
Wilmar, J. C., 53
Wilson, D., 170
Wilson, H. M., 137, 138, 139, 144
Wilson, P. D. G., 120
Windholz, M., 35, 53, 56
Withers, L. A., 72, 102, 197, 199
Wolter, K. E., 50
Wood, H. N., 7
Wright, K., 53
Wullems, G., 12

Yamada, Y., 70, 102, 120
Yamamoto, Y., 70
Yeoman, M. M., 44, 69, 70, 71, 205, 207, 216, 218

Zbell, B., 105
Zenk, M. H., 11, 206
Zerman, H., 105

# Subject index

ABA (abscisic acid), 47, 53, 197–8
Acacia, 118
acclimation period, see hardening off
Acer pseudoplatanus (sycamore), 94, 206
Acer rubrum, 118
acetocarmine staining (of chromosomes), 137, 145–6
actin, 158
actinomycin D, 187
adenine, 7, 47
adenine sulfate, 82, 128
ADP-ribosyltransferase, 158
aeration
  of root cultures, 117
  of suspension cultures, 94–5
African violet, see Saintpaulia ionantha Wendl.
agar
  contaminants, 44
  embedding of protoplasts, 175–6
  see also medium, matrix for
agar-plating method (Bergmann), 8
Agargel, 52
Agave, 70
Agrobacterium-mediated transformation, 12
Agrobacterium rhizogenes (strain, 15834; Ri plasmid), 85, 120, 124
Agrobacterium tumifaciens (crown gall), 6, 8, 10
alanine, 104
alcohol, acidified, 32
alfalfa, see Medicago sativa
alkaloid, 204–6; see also Ruta graveolens
allergen, 205
aluminum foil
  closures, 38
  grades, 28
aminoacetic acid, see glycine
amino acid
  and other nitrogenous medium additives, 49
  possible contaminants of sucrose, 46
  reactions with carbohydrates, 53

p-aminobenzoic acid, 45
1-α-aminooxy-β-phenylpropionic acid, as inhibitor of lignification, 158
ammonium chloride, and embryoid formation, 104
ammonium nitrate, and shoot formation, 54
Ananas cosmosus (pineapple), 106
ancymidol, 198
anergy, see habituation
anther culture
  anther tissue influences, 138–9
  categories of anthers (pre-/post- mitotic), 137
  charcoal in, 138
  chromosome numbers in, 139
  light requirements, 143–4
  physiological status of plants, 138
  possible developmental pathways of microspores, 140
  procedures, 142–6
  stage of development of anthers, 137
  treatment of anthers prior to, 138
anthocyanin pigment, 70, 208–9
anthraquinone, 207, see also Morinda citrifolia
antibiotic, 34–5, 204
antioxidant, 73, 130
aphidicolin, 157
apical meristem culture, 126–8
apple, see Malus domestica
Arabidopsis, 53
arabinose, 170
L-arginine, 49
arginine decarboxylase, 106
Armoracia lapathifolia (horseradish), 85
aromatic terpenes, 209
Artocarpus spp. (breadfruit, jackfruit), 195
ascorbic acid, 45
L-asparagine, 4, 49
Asparagus, 22
Asparagus officinalis, 126
L-aspartic acid, 49
Aspergillus niger, 27

*Atropa belladonna*, 102, 137, 167–8
autoclave, *see* wet-heat sterilization
auxin
    callus growth and, 71
    defined, 46
    embryogenesis and, 104
    habituation and, 10, 48
    micropropagation and, 128
    ratio to cytokinin, 7, 83, 93
    role in organogenesis, 82
    xylem differentiation and, 156
    *see also* 4-CPA; 2,4-D; dicamba; IAA; α-NAA; 2,4,5-T
*Avena sativum* (oat), 185
axenic germination, *see* surface sterilization, seeds

B5 medium, 54
B-995 (succinic acid-2,2-dimethyl hydrazide), 198
*Bacillus*, heat-resistant species, 30
bacteriological glove box, *see* laminar flow cabinet
bacteriostatic agent, 28
banana, *see* Musa spp.
banana powder, 50
BAP (6-benzylaminopurine), 7, 48, 54, 84, 104, 128–9
barley, *see* Hordeum vulgare
batch culture, 95
Bayoun disease, *see* Fusarium oxysporum Schlet. var. *albedinis*
*Begonia rex*, 84
6-benzylaminopurine, *see* BAP
Bertholot's salt mixture, 6
*Beta vulgaris* (sugar beet), 49, 148, 176, 198
bioreactor, 106
biotin, 45
boiling-water bath, 30
boring pathform, 76
*Brassica napus*, 141
*Brassica oleracea* var. Borytis, 102
*Brassica rapa* (turnip), 72
brassinosteroid, 156
bromeliad, 22
*Bromus inermis* Leyss. var. Manchar, 102
bud initiation, *see* shoot initiation

$C_3$, $C_4$ plants, 43, 169
calcium chloride (monohydrate) and membrane stability, 179
calcium hypochlorite, 32
calcium pantothenate, 45, 53
calcium *ortho*phosphate, mono (prim.) and membrane stability, 179
Calcofluor White M2R, 179
callose, 69
callus
    defined, 4, 69
    friability, 70–1, 93
    growth and development, 70–1

hormonal requirements, 71
plant material for, 72
procedure for initiation, 75–7
subculture, 71
*Canavalia ensiformis*, 43
*Candida albicans*, 34
carbol-fuchsin stain, 218
carbon
    adsorption, 51
    sources for media, 45
cardiac glycosides, 204
*Carica papaya*, 102
carnation, *see* Dianthus caryophyllous
carrot, *see* Daucus carota
casamino acids, 49
casein hydrolysate, 49, 137
cassava, *see* Manihot esculentum
*Catharanthus roseus*, 206
CCC ([2-chloroethyl]-trimethylammonium chloride), 198
cell count
    density, 94, 161, 215, 217, 219, 222–3
    total, 215
cell differentiation, *see* transdifferentiation; xylem cell differentiation
cell separation, *see* chromium trioxide
cellulase, 9; *see also* Onozuka R-10
Cellulysin, 178
cell wall
    composition, 170
    regeneration, 168, 179
centrifugation
    in anther pretreatment, 138
    in PCV determination, 216
    in protoplast fusion, 185
    in protoplast purification, 173–4
    in synchronization of embryogenesis, 105
    in viability determination of cell suspensions, 99
charcoal (activated)
    anther cultures and, 138
    embryogenesis stimulated by, 50, 104
    possible modes of action, 50, 138
    wood vs. bone, 50
chemical sterilization, 32
chemostat, 8, 95
chloramphenicol, 37
*Chlorella*, 216
chlorine (gas), 35
(2-chloroethyl)-trimethylammonium chloride, *see* CCC
*p*-chlorophenoxyacetic acid, *see* 4-CPA
choline chloride, 45
chromium trioxide (chromic acid), 98, 217
*Chrysanthemum*, 128
*Citrus*, 31, 49, 71, 169
*Citrus sinesis* var. "Shamouti" orange, 102
Cleansphere, 20
Clorox, 32; *see also* sodium hypochlorite

closed continuous culture, 95
closed culture, 95
cobalt, 43
coconut, see Cocos nucifera
coconut water, preparation, 55
Cocos nucifera (coconut), 6, 9, 48–9, 73, 79, 137, 195–6
Coffea, 106
Coffea arabica, 102
colchicine, 136, 139–40, 145
Coleus blumei, 126
Colocasia esculentum (taro), 195
complementation selection procedure, 187
Comptonia, 118
contamination
 avoidance, 37
 of tissue cultures, 34
copper oxidase (polyphenoloxidase; polyphenolase), 73, 130
cork borer, care of, 79
corn, see Zea mays L.
cotton, see Gossypium hirsutum
coumarate–coenzyme A ligase, transdifferentiation and, 158
coumarin derivative, see Ruta graveolens
4-CPA (p-chlorophenoxyacetic acid), 47, 54
Crassulacean acid metabolism, 43, 169
crown gall, see Agrobacterium tumifaciens
cryopreservation, 72, 197
cyanocobalamin, 45
cybrid (heteroplast; cytoplasmic hybrid), 9, 184
Cymbidium, 127
cysteine-HCl, 6
cytodifferentiation
 secondary-product biosynthesis and, 207
 tracheary elements produced by, 8
 see also xylem cell differentiation
cytokinin
 bud proliferation in shoot apex, 10
 callus growth and, 71
 defined, 7, 46
 embryogenesis and, 104
 habituation, 10, 48
 micropropagation and, 128
 ratio to auxin, 7, 82, 93
 thiamine and, 45
 xylem differentiation and, 156
 see also adenine; adenine sulfate; BAP; 2iP; kinetin; zeatin
cytoplasmic hybrid, see cybrid

2,4-D (2,4-dichlorophenoxyacetic acid)
 callus stimulated by, 6, 73
 component of media, 46–7, 54
 structural formula, 47
 suppresses organogenesis, 48, 128
 thermostability, 52
 transdifferentiation with, 157
Dactylis, 198
Dactylis glomerata (orchard grass), 105

Dahlia, 128
date palm, see Phoenix dactylifera
Datura, 6, 9
Datura innoxia, 136–7
Daucus carota (carrot)
 callus induction, 6, 72, 78
 cambial explants, 5, 69
 cultures: apical meristem, 126; form anthocyanin, 70, 208; protoplast, 176; stimulated by 2,4-D and coconut milk, 6; suspension, procedure, 97–8
 embryogenesis, 31, 101–2, 104, 107
 polyamines and embryogenesis, 106
 determination, 155
Dianthus caryophyllous (carnation), 128
dicamba (3,6-dichloro-O-anisic acid), 105
2,4-dichlorophenoxyacetic acid, see 2,4-D
6[$\gamma$,$\gamma$-dimethylallylamino]purine, see 2iP
dimethyl sulfoxide (DMSO), 32, 56
Dioscorea (yam), 195
Dioscorea deltoides, 206
diosgenin, 205–6; see also Dioscorea deltoides
1,3-diphenylurea, 48
dishwashing, laboratory requirements for, 20–1
disinfection, defined, 28
DNA, 7, 157, 167
 excision repair, 158
 see also T-DNA
L-dopa, 205
Douglas fir, see Pseudotsuga menziesii
Driselase, 178
dry-heat sterilization, 28
dry weight, determination, 215–16, 218, 221–2
Durapore filter, 32

EDTA, 44, 53, 217; see also ferric ethylenediamine tetraacetate, sodium salt of
Elaeis guineensis (oil palm), 196
electrodialysis, 51
embryogenesis, discovery, 7
embryogenic clump formation, 103
embryogenic potential, 105
embryoid (somatic embryoid)
 anther and pollen cultures produce, 137
 charcoal stimulates formation, 50, 104
 cytokinins and, 104
 initiation, 107
 nitrogen:auxin ratio, 104
 polyamines and, 106
 potassium and, 54
 procedure for induction, 107
 protoplast cultures form, 176
 separation of fractions, 108
 separation of stages, 105
 sources, 101
 stages in development, 102–3
 synchronization of formation, 105
embryo rescue, 110

English ivy, see Hedera helix
energy sink water reservoirs, see microwave sterilization
environment room, 22
epidermal/subepidermal explant, 83
Equisetum, 9
essential oil, 207
ethanol
  surface disinfection with, 26–7, 32
  two-stage disinfection with, 33–4
ethanol dip, 32, 35–6
ethylene, 37, 46, 49, 85, 156
ethylene oxide, 36
Eucalyptus grandis, 130
Euphorbia, 70
Evans blue dye, 161, 175
explant cutting guide, 79

face mask, 37
fermenter, industrial pilot plant, 96
fern, 9, 69, 126
ferric ethylenediamine tetraacetate, sodium salt of (NaFeEDTA), 44, 118
Festuca, 198
Feulgen stain, 145–6
filter, 25, 31, 32, 51, 98
  prevents light degradation, 53–4
filter paper
  bridge, 108, 128
  disk, 52
  platform, 52
  see also paper-raft technique
flaming instrument, 32
flavonoid, 207
flavonol, see Acer Pseudoplatanus
fluorescein diacetate, 175
fluorescence microscopy, 175
fluorescent cell sorting, 186
5-fluorodeoxyuridine, 157
Fluoropore filter, 32
folic acid, 45
formazan precipitate, 99
Fragaria vesca (strawberry), 31, 198
Fraxinus, 50
freeze preservation, see cryopreservation
fresh weight, determination, 215, 218, 220–1
friable culture, 70–1, 93
D-fructose, 45, 53
Fuchs–Rosenthal hemocytometer, 161, 175
6-furfurylaminopurine, see kinetin
Fusarium oxysporum Schlect. var. albedinis (Bayoun disease), 196
fusigenic agent, 184

galactose, 170
gas–liquid chromatography, 209
gas sterilization, see ethylene oxide
gelatine, 6
Gelrite, 59
gene gun, 12
gentamicin sulfate, 35

geranium, see Pelargonium
Gerbera, 22
germicidal (UV) lamp, 26
  health risks, 35–6
  monitoring effectiveness, 26
germplasm preservation
  conventional methods involve losses, 195
  procedures, 199–201
gibberellic acid ($GA_3$), 46, 128, 156
gibberellin, 46, 52, 85
Gingko biloba, 9, 136
ginseng saponin (ginseng saponin glycoside), 205; see also Panax ginseng
D-glucose, 6, 45, 53, 118
L-glutamic acid, 49, 104
L-glutamine, 49, 104
glycine (aminoacetic acid), 42, 49
Glycine max (soybean)
  callus initiation, 72
  nickel and, 43
glycyrrhizin, 205
Golgi cisternae, rosettes, 158
Gossypium hirsutum (cotton), 169
grape, see Vitis rupestris
Gro-Lux lamp, 22, 128
gymnosperm, 9, 69, 118

$H^3$-thymidine, see tritiated thymidine
habituation (anergy), 10, 48
hairy root, 85, 120, 124
halophyte culture, 43
Haplopappus gracilis, 45, 208
hardening off, 88, 129, 132
health hazard, 27, 35
Hedera helix (English ivy), 104
Helianthus tuberosus (Jerusalem artichoke; sunchoke), 35
hemicellulase, 170, 178
Hemerocallis, 178
hemocytometer, see Fuchs–Rosenthal hemocytometer; Sedgwick–Rafter slide
henbane, see Hyoscyamus niger
HEPA filter, 25
heterokaryon (heterokaryocyte), 10, 183–4
heteroplast, see cybrid
hexochlorophene, 27
high-temperature degradation of compounds, see wet-heat sterilization
Hirsh funnel, 98
homokaryon, 184
Hordeum vulgare (barley), 33, 102, 137
horseradish, see Armoracia lapathifolia
5-(hydroxymethyl)-2-furfural, 138
Hyoscyamus niger (henbane), 137, 139, 141, 146, 177

IAA (indole-3-acetic acid), 6, 47
  auxin definition, 46
  degradation, 47, 52–3
  micropropagation and, 128
  shoot-apex cultures and, 128

structural formula, 47
see also auxin
IAA-*myo*-inositol conjugate, 45
IAA oxidase, 47
IAPTC (International Association for Plant Tissue Culture), 24
IBA (indole-3-butyric acid), 47, 53, 84, 128–9
incubation of culture, laboratory requirements for, 21
indole-3-acetic acid, see IAA
indole-3-butryic acid, see IBA
*myo*-inositol, 45, 118
inositol bisphospholipid, 45
iodine, a possible micronutrient, 43
2iP (isopentylaminopurine; 6[$\gamma,\gamma$-dimethylallylamino]purine $N^6$-[$\Delta^2$-isopentyl]adenine), 7, 48, 52, 128
*Ipomoea batatas* (sweet potato), 72, 195, 198
iron stock (MS), preparation, 44, 56
isopentyl adenine, see 2iP
isopentylaminopurine, see 2iP
isopropanol, 26–7, 32

jackbean, see *Canavalia ensiformis*
Jerusalem artichoke, see *Helianthus tuberosus*

kinetin (6-furfurylaminopurine), 48
 anther cultures and, 137
 callus medium ingredient, 54, 73
 cytokinin definition, 46
 discovery, 7
 embryogenesis in cell suspensions, 104
 micropropagation and, 128
 structural formula, 47
 thermostability, 52
 transdifferentiation and, 157
Knop's solution, 4, 6

$\beta$-lactam, 37
*Lactuca sativa* (lettuce)
 callus initiation, 72
 gentamicin sulfate and, 35
 xylem cell formation in callus, 46, 162
laminar flow cabinet, 20, 25
latex, 207
lemon, see *Citrus*
leptohormone, 4
lettuce, see *Lactuca sativa*
light
 degradation of IAA and IBA, 47, 53
 for micropropagation, 128
 for organogenesis, 85
 photooxidation of EDTA, 53
 for plantlets from anther cultures, 144
 protoplast cultures and, 176, 179
 shoot-apex cultures and, 128
 yellow long-pass filters prevent degradation by, 53–4
lignin/lignification, 158, 207

*Lilium*, 85
linear growth rate, of root cultures, 215
liquid nitrogen, 201
liverwort, 69
*Lolium*, 198
low-temperature storage, of tissue culture, 198, 201
*Lycopersicon esculentum* (tomato)
 aeration of root cultures, 117
 anther cultures, 148
 protoplast fusion with potato, 12
 root cultures: grown at various sucrose concentrations, 119; nutritional requirements, 5, 42
Lysol, 36

Macerase, 178
maceration of cells, see chromium trioxide
Macerozyme R-10, 173, 178
macronutrient, essential element, 43
magnesium sulfate and membrane stability, 179
maize, see *Zea mays* L.
maleic hydrazide, 198, 200
malt extract, 49
*Malus domestica* (apple), 198
*Manihot esculentum* (cassava), 102, 195, 198
mannitol, 50, 169, 173, 197–8
*Medicago sativa* (alfalfa), 178
medium
 inaccuracies in formulation, 44
 laboratory requirements for preparation, 21
 matrix for, 51
 requirements for plant tissue culture, 42
 selection, 54
 see also specific medium
Meicelase (CESB, CMB), 178
*Mentha* (mint), 208
mercuric chloride, 36
MES (2-[$N$-morpholino]ethanesulfonic acid) buffer, 60
L-methionine, 49
*Micrococcus*, 34
microfiltration, 31
 pore size required, 32
micronutrient
 essential elements, 43
 preparation of MS stock, 56, 58
micropropagation
 procedures, 131–2
 stages, 127
 woody plants, 129–30
 see also bioreactor
microspore culture, see anther culture
microtubule, 158
microwave sterilization, 30–1
Millex filter, 31
minimal-growth storage, methods, 197
minimum limit, 4
mint, see *Mentha*

mitotic index, 215, 217, 219, 224–5
mixed-enzyme method of wall degradation, 170
monocot culture, 9, 84
*Morinda citrifolia*, 206
moss, 69
MS medium, preparation of complete, 58
MS stock, preparation, 56–8
*Musa* spp. (banana, plantains), 106, 195–6
mutagen, 11

α-NAA (naphthaleneacetic acid)
    callus growth and isomers of, 47
    micropropagation and, 128
    root initiation with, 84, 129
    structural formula, 47
    thermostability, 52
niacin, *see* nicotinic acid
nickel, possible micronutrient, 43
*Nicotiana glauca*, 6, 186
*Nicotiana langsdorffii*, 6, 186
*Nicotiana rustica*, 206
*Nicotiana tabacum* (tobacco)
    anther cultures, 137–8, 141
    apical meristem cultures, 126
    callus: initiation, 7, 72; medium, 69
    light-sensitive mutant, 187
    microspore cultures, 9
    organogenesis in cultures, 82, 85
    protoplast: fusion, 186–7; isolation, 169, 177 (L. var. Xanthi)
    somatic embryogenesis (var. Samsun), 102
    thiamine in cultures, 45
nicotine, *see Nicotiana rustica*
nicotinic acid (niacin), 42, 44, 53
nurse culture, *see* paper-raft technique
nylon mesh filter, 98

oat, *see Avena sativum*
oil palm, *see Elaeis guineensis*
Onozuka R-10, cellulase, 173, 178
oligosaccharide, 84
open continuous system, 95
orange juice, 49
orbital shaker, 21, 23
orchard grass, *see Dactylis glomerata*
orchid propagation, *see Cymbidium*
organ culture, vs. tissue culture, 5
organic supplement to medium, 49
organogenesis
    auxin:cytokinin ratio induces, 7
    culture procedure for, 87–8
    epidermal explants demonstrate, 83
    ethylene and, 85
    factors involved in, 82
    gibberellins and, 85
    oligosaccharides, 84
    organs initiated by, 82
    *see also* root initiation; shoot initiation
*Oryza sativa* (rice), 139
osmoticum (plasmolyticum), 50, 169

oven sterilization, *see* dry-heat sterilization
oxygen, *see* suspension culture, aeration
ozone toxicity, 27, 36

packed cell volume, 215–16, 219, 223–4
*Paeonia*, 137
*Panax ginseng*, 120, 206
*Papaver bracteatum*, 206
paper-raft technique, 4, 8
Parafilm M, 38, 89, 107, 144, 173
passage time, 71, 93
pathogen-free plant, 10–11
pea, *see Pisum sativum*
pectin, 170
pectinase, 9, 170, 178, 217
Pectolyase Y-23, 178
*Pelargonium* (geranium), 130, 169
peptone, 4, 49
*Persea americana* (avocado), 195
*Petunia*, 138, 169, 175, 177, 179
*Petunia hybrida*, 148, 187
*Petunia parodii*, 187
pH, important in medium preparation, 59–60
phenolic, 26, 36, 84, 130, 206; *see also Acer pseudoplatanus*
N-phenyl-N'-1,2,3-thiadiazol-5-ylurea, *see* TDZ
phenylpropanoid, 205
pHisohex, 27
*Phleum*, 198
phloroglucinol, 84
*Phoenix dactylifera* (date palm), 33, 196
Phosfon D (tributyl-2,4-dichlorobenzyl-phosphonium chloride), 198
phosphate and shoot formation, 54
phosphatidylinositol, 45
phosphoinositides, 45
photoautotropic culture, 45
Phytagel, 52
phytochrome, 85
phytosterol, *see* sterol
picloram (Tordon), 47–8, 73
pineapple, *see Ananas cosmosus*
*Pinus ponderosa* Dougl., 102
*Pisum sativum* L. (pea), 178
plant growth regulator
    vs. hormone, 46
    solvent for, 56
plasmid DNA, *see* T-DNA
plasmolyticum, *see* osmoticum
pollen basal medium, 145
pollen culture, 9, 144
    *see also* anther culture
polyamine, 106
polycarbonate culture flask, reaction of ethylene oxide with, 37
polyester fleece, 52
polyethylene container, prolonged water storage in, 51
polythylene glycol (PEG), 50, 185
polymerase, α and β types, 157

polyphenoloxidase (polyphenolase), see copper oxidase
polypropylene film, capping culture tubes with, 38–9
polyvinylpyrrolidone, 169
*Populus*, 148
 shoot initiation on stem explants of, 89
pore size, for microfiltration sterilization, 32
potassium nitrate, in embryogenic medium, 104
potato, see *Solanum tuberosum*
premixed culture medium, 10
pressure cooker, see wet-heat sterilization
proembryoid cell, 103
proteinase inhibitor, see *Scopolia japonica*
protoplast, 9
 agar embedding, 175–6
 culture conditions, 175–6, 178–9
 density, 175
 embryogenesis, 176
 foreign material introduced into, 167
 fusion, 12, 167, 186–7, 188–9; and surface charge, 185
 isolation technique, 169–70, 172–5
 medium preparation, 176
 membrane stability, 179
 regeneration: of cell wall, 168, 179; of plants, 176
 selection of plant material, 177
 viability, 175
 wall-degrading enzymes, 170, 177–8
*Prunus*, 196
*Pseudotsuga menziesii* (Douglas fir), 52
putrescine, 106
pyridoxine, 42, 44, 53

quantitative measurement
 cell count: density, 219, 222–3; total, 215
 dry vs. fresh weight, 218, 221–2
 linear growth rate, 215
 mitotic index, 219, 224–5
 packed cell volume, 219, 223–4
 tracheary element count, 220, 225–6

radioimmunoscreening method, 206
radish, see *Raphanus sativus*
*Ranunculus*, 102
*Ranunculus sceleratus*, 102
rape, see *Brassica napus*
*Raphanus sativus* (radish), 73, 118
reinvigoration, 129
rejuvenation, 129
resin, 204, 207
resin-based deionization, 51
reverse osmosis, 51
*Rhizopus nigricans*, 27
Rhozyme HP-150, 178
riboflavin, 45
riboside, 7
ribotide, 7

*Robinia*, 118
root culture, 5
 advantages, 117–18
 aeration, 117
 linear growth measurement, 214–15
 nutritional requirements, 118–19
 procedure, 122–3; for subculture, 120
 salinity and, 123
 sucrose concentrations and, 119
 see also *Lycopersicon esculentum*
root, intact vs. cultured, 119
root initiation, 7, 82, 84, 129; see also organogenesis
*Rosa*, 70
*Ruta graveolens*, 206
rye, see *Secale cereale*

*Saccharum officinarum* (sugarcane), 45, 102
safranin O, 162
*Saintpaulia ionantha* Wendl. (African violet), 85
salinity, and root growth, 123
saponin, 204; see also ginseng saponin
*Saxifraga*, 22
scanning electron microscopy, 146
Schenk and Hildebrandt medium, 105
*Scopolia japonica*, 206
*Scorzonera hispanica*, 10
screen and depth filters, 51
SeaPlaque, 52
*Secale cereale* (rye), 141
secondary metabolism, 12, 70, 204
secondary-metabolite production, procedure, 209
sector inoculum, 120
Sedgwick–Rafter slide, 98, 217, 226
*Selaginella*, 9
sequential enzyme method, of wall degradation, 170
*Sequoia sempervirens*, 9
serpentine, see *Catharanthus roseus*
shoot-apex culture, 127–8
shoot-apex technique (Morel), 10
shoot-induction medium, 54, 84
shoot initiation, 7, 82, 84; see also organogenesis
sodium
 possible micronutrient, 43
 toxicity, 33
sodium hypochlorite, 32
 buffering, 33
sodium nitrate, fusigenic effect, 185
*Solanum lacinatium*, 206
*Solanum melongena* L., 102
*Solanum tuberosum* (potato)
 apical and shoot meristem cultures, 128
 callus cultures, 6, 72–3
 embryo rescue, 40
 freeze preservation of shoot apices, 201
 germplasm: collection, 195; conservation, 198

## Subject index 255

micropropagation, 130
protoplasts, 11, 12, 169, 178
surface sterilization of tuber, 33
solasadine, see *Solanum lacinatium*
solvent, for plant growth regulator, 56
somatic embryogenesis, see embryoid
sorbitol, 50, 169, 197–8
soybean, see *Glycine max*
spermidine, 106
spermine, 106
stage-2 (stage-4), 137
*Staphylococcus*, 34
*Stellaria media*, 126
sterilization, defined, 27–8; see also specific type
sterol, 204, 207
stock preparation, 87
 iron stock (MS), 44, 56
 preparation of MS stock, 56, 58
stock solution, preparation, 55
strawberry, see *Fragaria vesca*
subculture, 71, 120; see also passage time
succinic acid-2,2-dimethyl hydrazine, see B995
sucrose
 nutritional component of media, 4, 42, 45, 118
 osmoticum, 169
 thermostability, 30, 53
 xylogenesis and, 46
sugar beet, see *Beta vulgaris*
sugarcane, see *Saccharum officinarum*
sunchoke, see *Helianthus tuberosus*
Sunderland and Roberts' medium, 138
surface sterilization
 seeds, 27, 33
 small and hydrophobic plant materials, 39
 tomato fruits, 122
 tubers, 33
 two-stage, 33–4
suspension culture
 aeration, 94–5
 cell density, 94
 cell-line characteristics, 92
 *Daucus carota*, procedure for, 97–8
 friability, 93
 growth stages, 93
 initiation, 7, 92
 passage time, 93
 viability, 99
sweet potato, see *Ipomoea batatas*
Swinnex filter, 31
Swinney filter, 31
sycamore, see *Acer pseudoplatanus*
synchronous division, 217
synkaryocyte, 183–4

2,4,5-T (2,4,5-trichlorophenoxyacetic acid), 47
tannin, 204, 207

taro, see *Colocasia esculentum*
*Taxus*, 9
T-DNA, 12, 120
TDZ (*N*-phenyl-*N*′-1,2,3-thiadiazol-5-ylurea; thidiazuron), 48
tetrazolium staining, see 2,3,5-triphenyltetrazolium chloride
thebain, see *Papaver bracteatum*
thiamine-HCl, 6, 42, 44, 53–4
*Theobroma cacao* (cocoa), 195
tissue culture
 defined, 5
 laboratory functions, 19
 see also specific type
tobacco, see *Nicotiana tabacum*
tocopherol, 45
toluidine blue O, 78
tomato, see *Lycopersicon esculentum*
tomato juice, 49
Tordon, see picloram
totipotency, 3, 7–8
tracheary element, 8
 count, 215, 220, 225–6
 see also xylem, anatomy; xylem cell differentiation
transdifferentiation, 155
 see also xylem cell differentiation
transfer chamber, see laminar flow cabinet
Transfergel, 52
transformation, 12
tributyl-2,4-dichlorobenzylphosphonium chloride, see Phosfon D
2,4,5-trichlorophenoxyacetic acid, see 2,4,5-T
2,3,5-triphenyltetrazolium chloride, 99, 102
tritiated thymidine ($H^3$-thymidine), 218
*Triticum aestivum* (wheat), 148
Triton, 33
*Tropaeolum majus*, 126
tumor tissue, tobacco, 5–6
tumor-inducing principle, see T-DNA
turbidostat, 8, 95
turnip, see *Brassica rapa*
Tween-20, -80, 33
L-tyrosine, 49

ultrafiltration, water purification by, 51
ultraviolet irradiation, see germicidal lamp
urea, 104
urease, 43
Uvcide Germicidal Lamp Monitor, 26

Vancomycin, 37
vascular nodule (meristemoid), 70
vitamin
 nutritional components of media, 44
 preparation of MS stock, 58
 thermostability, 53
*Vitis rupestris* (grape), 198
volatile oil, 204

washing of hands, prior to culture, 27
water-purification system, 51
wet-heat sterilization, 28
  degradation of media components by, 30, 45, 52–3
  time requirement, 29
wheat, *see Triticum aestivum*
Whipple disk/field, 223, 226
White's medium, 121–2
working area, selection, 25
wound callus, 3, 69
wound hormone, 4

xylan, 158, 170
xylem, anatomy, 156
xylem cell differentiation
  advantages in studying, 155
  cell cycle and, 157
  cultural requirements, 156
  ethylene and, 46, 49, 156
  gentamicin sulfate inhibits, 35
  induction, 156, 162
  medium for, 156
  microtubules and, 158
  secondary-wall patterns, 158
  sucrose requirement, 46
  quantitation, 225–6
xylem fiber, 156

yam, *see Dioscorea*
yeast extract, 49, 137
yield vs. growth, 214

*Zea mays* L. (maize), 148, 185
zeatin, 7, 47–8, 52, 84, 104, 128
zeatin riboside, 7
*Zinnia*
  cell density, 161
  isolation of mesophyll cells, 160
  leaf disks, 163
*Zinnia elegans*, 155, 157–8

Printed in the United States
By Bookmasters